AGRICULTURE ISSUES AND POLICIES

MILLION DOLLAR FARMS IN THE NEW CENTURY

AGRICULTURE ISSUES AND POLICIES

Additional books in this series can be found on Nova's website at:

https://www.novapublishers.com/catalog/index.php?cPath=23_29&seriesp=Agriculture+Issues+and+Policies

Additional e-books in this series can be found on Nova's website at:

https://www.novapublishers.com/catalog/index.php?cPath=23_29&seriespe=Agriculture+Issues+and+Policies

AGRICULTURE ISSUES AND POLICIES

MILLION DOLLAR FARMS IN THE NEW CENTURY

SAMUEL D. BOSWORTH
EDITOR

Nova Science Publishers, Inc.
New York

Copyright © 2010 by Nova Science Publishers, Inc.

All rights reserved. No part of this book may be reproduced, stored in a retrieval system or transmitted in any form or by any means: electronic, electrostatic, magnetic, tape, mechanical photocopying, recording or otherwise without the written permission of the Publisher.

For permission to use material from this book please contact us:
Telephone 631-231-7269; Fax 631-231-8175
Web Site: http://www.novapublishers.com

NOTICE TO THE READER

The Publisher has taken reasonable care in the preparation of this book, but makes no expressed or implied warranty of any kind and assumes no responsibility for any errors or omissions. No liability is assumed for incidental or consequential damages in connection with or arising out of information contained in this book. The Publisher shall not be liable for any special, consequential, or exemplary damages resulting, in whole or in part, from the readers' use of, or reliance upon, this material. Any parts of this book based on government reports are so indicated and copyright is claimed for those parts to the extent applicable to compilations of such works.

Independent verification should be sought for any data, advice or recommendations contained in this book. In addition, no responsibility is assumed by the publisher for any injury and/or damage to persons or property arising from any methods, products, instructions, ideas or otherwise contained in this publication.

This publication is designed to provide accurate and authoritative information with regard to the subject matter covered herein. It is sold with the clear understanding that the Publisher is not engaged in rendering legal or any other professional services. If legal or any other expert assistance is required, the services of a competent person should be sought. FROM A DECLARATION OF PARTICIPANTS JOINTLY ADOPTED BY A COMMITTEE OF THE AMERICAN BAR ASSOCIATION AND A COMMITTEE OF PUBLISHERS.

LIBRARY OF CONGRESS CATALOGING-IN-PUBLICATION DATA

ISBN : 978-1-60741-755-2

Available Upon Request

Published by Nova Science Publishers, Inc. ✝ New York

CONTENTS

Preface		**vii**
Chapter 1	Million-Dollar Farms in the New Century *Robert A. Hoppe, Penni Korb and David E. Banker*	1
Chapter 2	United States Farm Income *Randy Schnepf*	53
Chapter Sources		73
Index		75

PREFACE

By 2006, million dollar farms, accounting for 2 percent of all U.S. farms, dominated U.S. production of high-value crops, milk, hogs, poultry, and beef. The shift to million-dollar farms is likely to continue because they tend to be more profitable than smaller farms, giving them a competitive advantage. Most million dollar farms (84 percent) are family farms, that is, the farm operator and relatives of the operator own the business. This book examines the growth of production from million-dollar farms since the 1980's and explores the current role of these farms in U.S. commercial agriculture, including their share of farms, their production of specific commodities, and their receipt of Government payments.

Chapter 1 - Million-dollar farms—those with annual sales of at least $1 million—accounted for about half of U.S. farm sales in 2002, up from a fourth in 1982 (with sales measured in constant 2002 dollars). By 2006, million-dollar farms, accounting for 2 percent of all U.S. farms, dominated U.S. production of high-value crops, milk, hogs, poultry, and beef. The shift to million-dollar farms is likely to continue because they tend to be more profitable than smaller farms, giving them a competitive advantage. Most million-dollar farms (84 percent) are family farms, that is, the farm operator and relatives of the operator own the business. The million-dollar farms organized as nonfamily corporations tend to have no more than 10 stockholders.

Chapter 2 - Despite high production costs, 2008 represented another year of record profitability for the U.S. farm economy as a whole. According to USDA's Economic Research Service (ERS), national net farm income—a key indicator of U.S. farm well-being—rose to a record $86.9 billion in 2008, marginally above the previous year's record ($86.8 billion). The growth in cash receipts to a record $323.4 billion for crop and livestock sales (up $38.6 billion or 14% from 2007) was nearly offset by a surge in production costs (up $38.2 billion or 15%) to a record $292.5 billion.

Government farm payments are projected up slightly in 2008 at $12.5 billion. An increase in ad hoc disaster payments more than offset lower commodity program payments, which declined when high crop prices rose above the price triggers for marketing loan benefits and counter- cyclical payments in 2008.

Within the farm balance sheet, total farm asset value of $2,359 billion and total farm debt of $212 billion are both projected at record levels in 2008. The debt-to-asset ratio of 9.0% is down sharply from last year's value of 9.6% and represents the lowest level since 1960, suggesting a strong financial position for the agricultural sector as a whole.

However, less than ideal market conditions heading into 2009 suggest dim prospects for the longer-term farm income outlook, albeit surrounded by considerable uncertainty. On the one hand, the global financial crisis, economic recession, rising unemployment, limited credit availability, and plummeting asset values that persist in early 2009 have contributed to substantial "demand destruction" (i.e., a severe weakening of consumer demand), which bodes poorly for farm commodity price prospects. On the other hand, weak energy markets and declining input prices could provide some spark to both producer investment and consumer demand for agricultural sector products, perhaps by the middle to latter half of the year. USDA will release its first U.S. farm income forecasts for 2009 on February 12, 2009.

This report supersedes CRS Report RS2 1970, *The U.S. Farm Economy*, by Randy Schnepf. It will be updated as events warrant.

In: Million Dollar Farms in the New Century ISBN: 978-1-60741-755-2
Editors: Samuel D. Bosworth © 2010 Nova Science Publishers, Inc.

Chapter 1

MILLION-DOLLAR FARMS IN THE NEW CENTURY

Robert A. Hoppe, Penni Korb and David E. Banker

ABSTRACT

Million-dollar farms—those with annual sales of at least $1 million—accounted for about half of U.S. farm sales in 2002, up from a fourth in 1982 (with sales measured in constant 2002 dollars). By 2006, million-dollar farms, accounting for 2 percent of all U.S. farms, dominated U.S. production of high-value crops, milk, hogs, poultry, and beef. The shift to million-dollar farms is likely to continue because they tend to be more profitable than smaller farms, giving them a competitive advantage. Most million-dollar farms (84 percent) are family farms, that is, the farm operator and relatives of the operator own the business. The million-dollar farms organized as nonfamily corporations tend to have no more than 10 stockholders.

Keywords: Contracting, family farms, farm businesses, farm financial performance, farm-operator household income, farm operators, farm structure, farm type, million- dollar farms

ACKNOWLEDGMENTS

The authors thank Hisham El-Osta, David H. Harrington, David A. McGranahan, James M. MacDonald, Mitchell Morehart, and Patrick Sullivan of the Economic Research Service (ERS), Michael D. Duffy of Iowa State University, Allan W. Gray of Purdue University, and Steven R. Koenig of the Farm Services Agency for their reviews and helpful comments. We also received editorial support, report design, and useful comments from Angela Anderson of the ERS Information Services Division.

SUMMARY

Small farms (those with annual sales less than $250,000) represent a large majority of U.S. farms (92 percent), but account for a relatively small share of total farm production (23 percent). This report examines the other end of the size spectrum, where a large percentage of farm production occurs, specifically on "million-dollar farms" whose annual sales total $1 million or more. The 35,100 million-dollar farms reported in 2006—2 percent of all U.S. farms—accounted for 48 percent of the sales of U.S. agricultural products.

What Is the Issue?

Understanding million-dollar farms is especially important because of the large and growing share of U.S. food and fiber they produce. This report examines the growth of production from million-dollar farms since the 1980s. It lays out the current role of million-dollar farms in U.S. commercial agriculture, including their share of farms, their production of specific commodities, and their receipt of Government payments.

What Did the Study Find?

Major shifts occurred in the distribution of gross farm sales between the 1982 and 2002 Censuses of Agriculture, with sales measured in constant 2002 dollars. Farms with sales of $1 million or more doubled their share of total U.S. farm sales from 23 percent in 1982 to 48 percent in 2002. Some of these million-dollar farms are relatively recent entrants to farming, while others existed as far back as 1978.

The shift in production to million-dollar farms is likely to continue. Average operating profit margins increase with sales, reflecting economies of size in farming. As a result, million-dollar farms—and farms growing to that size—have a competitive advantage relative to smaller farms. The shift in production may eventually slow, however, once million-dollar farms' shares of the commodities most amenable to large-scale production reach their upper limits.

Million-dollar farms do not have market power. The shift in farm production to million-dollar farms reflects a long-term concentration of farm production on fewer farms that has been underway since the beginning of the 20th century. However, there are still too many million-dollar farms—just over 35,000—for any single farm to dominate agriculture or the production of specific commodities.

Million-dollar farms receive a small share of Government payments. Most Government payments are commodity-related or targeted at current or past production of specific commodities, largely feed and food grains, cotton, and oilseeds. Relatively few million-dollar farms—particularly those with sales of $5 million or more—specialize in crops covered by commodity programs. As a result, million-dollar farms received only 16 percent of U.S. Government payments in 2006, a small share compared with their 48-percent share of gross sales, although disproportionately large compared with their 2-percent share of all farms.

Million-dollar farms have more operators than farms with lower sales. The number of operators per farm averaged 1.5 for all farms in 2006, but 2.1 for all million-dollar farms and 2.6 for $5-million farms. Multiple-operator farms accounted for 66 percent of million-dollar farms, substantially more than the 46-percent share for farms in general. Multiple-generation farms— those with at least 20 years' difference between the ages of the oldest and youngest operators—made up a larger share of million-dollar farms (23 percent) than any other sales class.

Most million-dollar farms are family operations. Eighty-four percent of the million-dollar farms in 2006 operated as family farms—defined as any farm where the majority of the business is owned by the operator (or the principal operator on multiple-operator farms) and individuals related to the operator. Only 7 percent of million-dollar farms were organized as nonfamily corporations, generally with no more than 10 stockholders. The situation was similar for farms with sales of $5 million or more, although a smaller share (64 percent) was classified as family operations and a larger share (21 percent) as nonfamily corporations. The operators and spouses on million-dollar farms, however, provided only 10 percent of the farms' labor, compared with 39 percent for farms with sales between $500,000 and $999,999.

Million-dollar farms account for most contract production. Thirty-nine percent of U.S. farm production came from farms with production or marketing contracts in 2006, and million-dollar farms accounted for about 62 percent of this contract production. Sixty-three percent of million-dollar farms used contracts, and about half of their production—mostly livestock— was under contract. Note that farms with production contracts only receive a fee from contractors, and only the fee—rather than sales—is included in their gross cash income. Measuring size by gross cash income rather than sales would reduce the number of million-dollar farms among some specializations, such as poultry farms.

Million-dollar farms also served as contractors. Approximately 5,400 farms reported contracting livestock production (including poultry) out to other farms. The share of farms contracting livestock production out was highest for $5-million farms at 12 percent.

How Was the Study Conducted?

Most of the data in this report are from the 2006 Agricultural Resource Management Survey (ARMS). The ARMS is a detailed, annual survey of farm businesses and associated households conducted jointly by the U.S. Department of Agriculture's Economic Research Service (ERS) and National Agricultural Statistics Service (NASS). The report also uses data from the last five censuses of agriculture to follow the shift in production to million- dollar farms. Finally, the 2002 Census of Agriculture Longitudinal File— which links records for individual farms from the last six censuses—traces the history of today's million-dollar farms.

INTRODUCTION

Between 1982 and 2002, the number of large farms—those selling at least $250,000 in farm products, measured in constant 2002 dollars—nearly doubled. The number of farms with sales of $1 million or more grew even faster, tripling over the same period, even after adjustments for inflation are considered. By 2006, 35,100 "million-dollar farms" accounted for 48 percent of U.S. agricultural sales. These farms use different management and marketing strategies than smaller farms. Half of their production occurs under marketing or production contracts, and two-thirds have more than one operator, generally not the spouse of the principal operator. Nevertheless, most million-dollar farms are family farms. Sixteen percent are classified as nonfamily farms.

This report examines data from the last five censuses of agriculture to follow the shift in production to million-dollar farms. We also used the 2002 Census of Agriculture Longitudinal File—which links records for individual farms from the last six censuses—to trace the history of million-dollar farms. Some million-dollar farms are relatively recent entrants to farming, while others go back as far as 1978.

Most of the data in this report are from the 2006 Agricultural Resource Management Survey (ARMS). The ARMS is a detailed, annual survey of farm business and associated households conducted jointly by the U.S. Department Agriculture's Economic Research Service (ERS) and National Agricultural Statistics Service (NASS).[1] Using ARMS data, the report presents a detailed examination of million-dollar farms, focusing on their:

- Share of farms, farm production, and Government payments;
- Specialization, farm size, and tenure;
- Business organization, including relatively new limited liability companies (LLCs);
- Operator characteristics, including the number of operators per farm;
- Farm business and farm household finances; and
- Farm business arrangements, including the use of hired labor and contracting.

Both the census and ARMS data used here were collected prior to the recent volatility in grain prices. The specializations of million-dollar farms may have changed somewhat as grain prices spiked and then fell in 2007 and

2008. For example, there may be more cash grain farms and slightly fewer livestock farms. It is unlikely that recent volatility in grain prices affected the overall conclusions of the report, the shift in production to larger farms, or the basic characteristics of million-dollar farms.

Recent ERS research on farm structure has focused on small farms, defined as those with sales less than $250,000 (USDA, National Commission on Small Farms, 1998). While most U.S. farms are small farms (92 percent, as of 2006), they account for only 23 percent of total U.S. farm production. This report examines the other end of the size spectrum, where a very large percentage of production occurs. Farms can sell a million dollars of products with a variety of enterprises, some of which do not require large amounts of farmland. For examples of million-dollar farms, see the box below.

A MILLION DOLLARS IN SALES

A variety of farm enterprises can generate $1 million in sales if they operate on a large enough scale. Some examples for different parts of the country are presented here. Estimates of the amount of crops and livestock necessary to generate $1 million of sales in 2005 are based on yields, livestock weights, milk production, and prices—generally reported at the State level—from *Agricultural Statistics 2007* (USDA, NASS, 2007).

Iowa corn and soybean farm, with a feedlot	Mississippi cotton and soybean farm	Arkansas broiler farm
Most of this farm's sales came from corn and soybeans, but it also supports a feedlot. The corn and soybeans shown below exclude any crops grown to feed the cattle.	Upland cotton (and cotton seed) accounted for more than 80 percent of the sales of this farm. However, the farm also harvested 960 acres planted to soybeans.	To generate just over $1 million in sales, this farm produced 475,000 broilers under production contracts. The broilers are actually owned by the contractor, who also provides specific inputs, such as feed. The farm operator receives a fee for his or her services provided. Thus, the gross revenue of the farm—largely the fees received—is much lower than the $1 million in sales, which includes the value of the birds removed by the contractor (Hoppe et al., 2007).
Sales (dollars) Corn 1,120 acres 375,894 Soybeans 800 acres 237,720 Fed cattle 400 head 448,656 **Total sales** **1,062,270**	Sales (dollars) Cotton 2,080 acres 823,678 Cotton seed 1,265 tons from the land in cotton 106,276 Soybeans 960 acres 198,326 **Total sales** **1,128,280**	

North Carolina hog farm	Wisconsin dairy farm	California specialty crops
Very large hog farms are a relatively recent development in North Carolina. Their introduction was facilitated by the use of contracts (McBride and Key, 2003). To generate $1 million in sales, the farm—a finish-only operation—produces 8,000 hogs weighing 256 pounds, the State average. No other farm enterprises are included in this example, since hogs account for 92 percent of sales of North Carolina hog farms, according to the 2002 Census of Agriculture.	A Wisconsin dairy farm milking 400 cows would generate slightly more than $1 million, calculated from average production and price estimates for the State. This example assumes no other farm products are sold, although feed for the cows could also be grown. According to the 2002 Census of Agriculture, 353 million-dollar dairy farms existed in Wisconsin. They accounted for 52 percent of all million-dollar farms in the State and 9 percent of million-dollar dairy farms in the United States.	California's specialization in high-value specialty crops dates back to the late 1800s (Cochrane, 1993). These examples show that even small acreages of specialty crops can generate sales of $1 million. Acres needed for sales of $1 million Head lettuce 170 Fresh tomatoes 125 Celery 120 Strawberries 35

LONGRUN TRENDS

Dramatic shifts have occurred in the distribution of gross farm sales since 1982. Farms with sales of $250,000 or more (in 2002 dollars) steadily increased their share of total sales from 47 percent in 1982 to 76 percent in 2002 (Figure 1). Farms with sales of $1 million to $4,999,999 and farms with sales of $5 million or more each doubled their share of sales between 1982 and 2002. Each of these sales classes accounted for nearly a fourth of agricultural sales by 2002.[2] Farms with sales just under $1 million—those selling $500,000 to $999,999—were the only other group to increase its share of sales over the period.

Number of Million-Dollar Farms

The number of million-dollar farms—less than 2 percent of U.S. farms in each census—tripled between 1982 and 2002 (table 1). The rate of increase for other large farms was not as high, but their numbers increased as well, especially those in the $500,000-$999,999 sales class.

The number of small farms declined, with the exception of "point farms" or farms with sales less than $1,000 that might normally have sales that high and satisfy the criteria to be counted as a farm.[3] Much of the increase in point farms, however, was due to an adjustment for undercoverage in the census farm count instituted in the 2002 census, which had the largest impact on farms near the $1,000 cut-off in farm definition (Allen, 2004; USDA, NASS, 2004). Adjusting the 1982 count of point farms for undercoverage, using published adjustment factors (U.S. Department of Commerce, 1985), reduced the 1982-2002 growth in point farms from 125 percent to 60 percent.

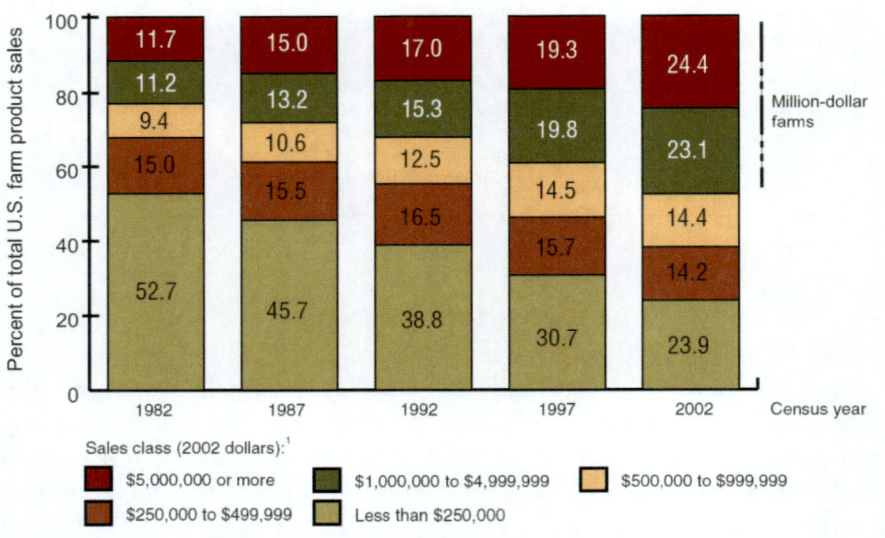

Note: Items may not add to totals due to rounding.
[1] Sales class is expressed in constant 2002 dollars, using the Producer Price Index for farm products to adjust for price changes.
Source: USDA, Economic Research Service, compiled from census of agriculture data.

Figure 1. Distribution of gross farm sales, by constant-dollar sales class,[1] 1982-2002 Million-dollar farms' share of sales increased from 23 percent in 1982 to 48 percent in 2002

Table 1. Number of farms, by constant-dollar sales class (2002 dollars),[1] 1982-2002

Sales class[1]	Census year					Change, 1982-2002
	1982	1987	1992	1997	2002	
	Number of farms					*Percent*
Total farms	2,240,976	2,087,759	1,925,300	1,911,859	2,128,982	-5.0
Small farms (sales less than $250,000)	2,156,057	1,989,883	1,807,605	1,775,875	1,976,646	-8.3
Point farms[2]	254,097	235,562	212,580	277,248	570,919	124.7
Other farms	1,901,960	1,754,321	1,595,025	1,498,627	1,405,727	-26.1
Large farms	84,919	97,876	117,695	135,984	152,336	79.4
$250,000-$499,999	57,691	64,195	74,354	78,330	81,694	41.6
$500,000-$999,999	18,242	22,058	28,583	36,469	41,969	130.1
Million-dollar farms	8,986	11,623	14,758	21,185	28,673	219.1
$1,000,000-$4,999,999	7,942	10,220	13,026	18,834	25,335	219.0
$5,000,000 or more	1,044	1,403	1,732	2,351	3,338	219.7

[1] Sales class is expressed in constant 2002 dollars, using the Producer Price Index for farm products to adjust for price changes. Point farms, however, are defined here in current dollars–rather than constant dollars–because they are identified in each census based on constant dollars.

[2] Point farms have sales of less than $1,000 (current dollars), but are still considered farms because they would be expected to normally sell at least $1,000 of agricultural products. In the 1997 and 2002 censuses, point farms included any establishments where sales of agricultural products and Government payments were less than $1,000. In this table, however, point farms are defined consistently from 1982 to 2002 as farms with sales less than $1,000, with no consideration of Government payments.

Source: USDA, Economic Research Service, compiled from census of agriculture data.

The History of Million-Dollar Farms

Census data—specifically the 2002 Census of Agriculture Longitudinal File—can be used to examine the history of million-dollar farms that existed in 2002. The longitudinal file links together the 1978, 1982, 1987, 1992, 1997, and 2002 Censuses of Agriculture, allowing analysts to track individual farms over the 24-year period.

Note that the census longitudinal file is not truly longitudinal. Rather than identifying farms and following them as time progresses, the longitudinal file links data collected in the past for another purpose: the agricultural census, which has its own issues regarding nonresponse and undercoverage. Because the census of agriculture was not designed to track businesses over time, errors linking records in the longitudinal file may lead to an overstatement of exits and entrances and an understatement of farms that remain in business. For more information about the Longitudinal File and its limitations, see, "Appendix: The 2002 Census of Agriculture Longitudinal File."

Table 2. Business age of farms, 2002

	All farms	Distribution of farms by business age						Total
		Less than 5 years[1]	5 to 9 years[2]	10 to 14 years[3]	15 to 19 years[4]	20 to 23 years[5]	24 years or more[6]	
	Number				*Percent*			
Total farms	2,128,982	36.8	22.1	11.0	8.2	5.5	16.3	100.0
Small farms (sales less than $250,000)	1,976,646	38.0	22.3	10.8	7.8	5.2	15.9	100.0
Point farms[7]	570,919	53.8	22.7	8.0	4.9	2.9	7.7	100.0
Other farms	1,405,727	31.6	22.1	11.9	9.1	6.2	19.2	100.0
Large farms	152,336	21.8	20.3	14.3	13.1	9.0	21.5	100.0
$250,000-$499,999	81,694	20.7	18.9	13.9	13.2	9.5	23.8	100.0
$500,000-$999,999	41,969	21.6	20.8	14.9	13.4	9.1	20.3	100.0
Million-dollar farms	28,673	25.1	23.4	14.6	12.6	7.6	16.6	100.0
$1,000,000-$4,999,999	25,335	24.4	23.4	14.8	12.8	7.9	16.8	100.0
$5,000,000 or more	3,338	30.6	24.0	13.7	11.1	6.1	14.6	100.0

Note: Items may not add to totals due to rounding.
[1] First appeared in the 2002 census. Entered between 1998 and 2002.
[2] First appeared in the 1997 census. Entered between 1993 and 1997.
[3] First appeared in the 1992 census. Entered between 1988 and 1992.
[4] First appeared in the 1987 census. Entered between 1983 and 1987.
[5] First appeared in the 1982 census. Entered between 1979 and 1982.
[6] First appeared in the 1978 census. Entered in 1978 or earlier.
[7] Point farms have sales of less than $1,000 (current dollars), but are still considered farms because they would be expected to normally sell at least $1,000 of agricultural products.
Source: USDA, Economic Research Service, compiled from 2002 Census of Agriculture Longitudinal File.

Business Age

Million-dollar farms are younger than other large farms (table 2).[4] Only 17 percent of all million-dollar farms have an estimated business age of at least 24 years, compared with 22 percent for large farms in general, a 5-percentage-point difference. At the other end of the business-age continuum, 25 percent of million-dollar farms are new establishments, estimated to be less than 5 years old, compared with 22 percent of all large farms. In addition, $5-million farms are even newer to the industry.

It may seem surprising that 25 percent of all million-dollar farms and 30 percent of $5-million farms that existed in 2002 could have entered farming no earlier than 1998 and still have sales at those levels. The large share of farms entering no earlier than 1998 may partially reflect difficulties linking census records over time. Nevertheless, earlier analyses of longitudinal data based on the census of agriculture established that farming—like other businesses—has high turnover, with thousands of businesses entering and leaving the sector each year (Hoppe and Korb, 2006; MacDonald et al., 2007).

Not all new million-dollar farms are startups, however. The census counts multiple locations of a farm business as individual farms, if the locations are operated separately or if they are located in different counties or States (USDA, NASS, 2007). Thus, if a large broiler farm (or integrator) expands by adding houses at a different location that operates as a separate business—perhaps with a new partner—the new location is counted as an entry. Two existing large farms could also combine to form a million-dollar farm.

The organization of new million-dollar farms suggests that they often draw on resources from more than one individual, since sole proprietorships are uncommon among these farms. For example, only 18 percent of $5-million entrants were organized as sole proprietorships. Another 27 percent were organized as partnerships, virtually all of them formal partnerships (registered under State law). Fifty-three percent were incorporated, either as family corporations (66 percent of new corporations) or as nonfamily corporations (34 percent). Approximately 59 percent of the new nonfamily corporations had more than 10 stockholders, as did 18 percent of the family corporations.

Following Farms Back through Time

It is possible to trace the history of individual farms. Figures 2 and 3 distribute farms in each of the million-dollar sales classes in 2002 by their constant-dollar sales for the previous census years. Forty-nine percent of farms with gross sales between $1 million and $4,999,999 in 2002 that also existed in 1997 had sales in the same range in 1997. The percentage dropped as we looked further back in time (figure 2). Only 10 percent of the farms from 2002 that existed in 1978 had sales of at least $1 million that year, while 50 percent had less than $250,000 in sales.

In contrast, most farms that existed in previous years with sales of $5 million or more in 2002 had at least $1 million in gross sales in the earlier years (figure 3). The smaller million-dollar farms (sales from $1 million to $4,999,999) that existed before 2002 appear to have grown into the $1-million class over time, while most $5-million farms sold at least a million dollars of products from the beginning.

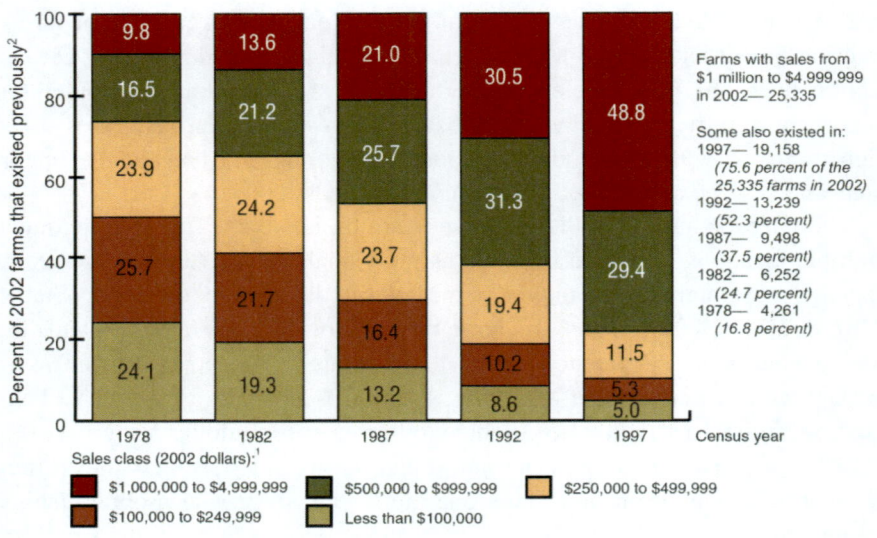

Note: Items may not add to totals due to rounding.
[1]Sales class is expressed in constant 2002 dollars, using the Producer Price Index for farm products. [2]Distributions are based on the number of farms in 2002 that also existed in previous years. For example, the first bar shows the distribution of the 4,261 farms in 2002 that also existed in 1978.
Source: USDA, Economic Research Service, compiled from the 2002 Census of Agriculture Longitudinal File.

Figure 2. Farms with sales of $1 million to $4,999,999 in 2002, by constant-dollar sales class,[1] 1978-1997
The share with sales of $1 million or more falls off rapidly in earlier years

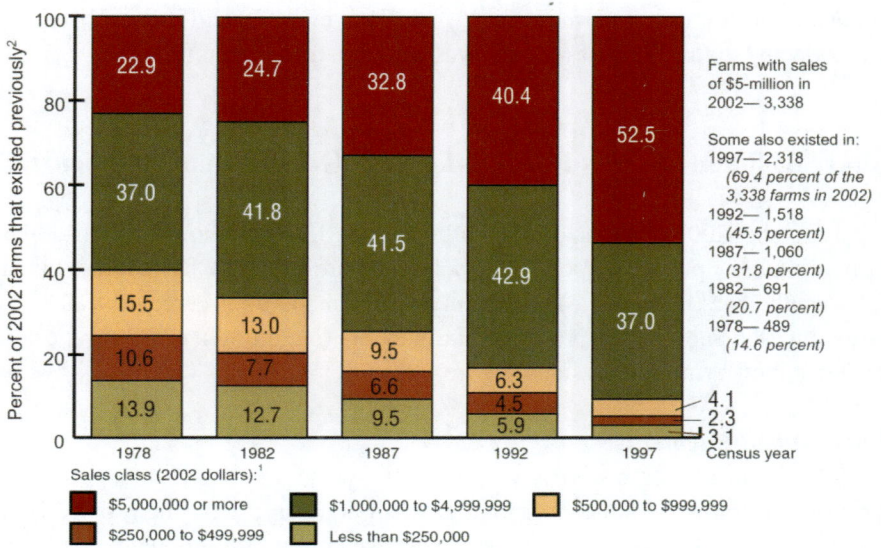

Note: Items may not add to totals due to rounding.
[1] Sales class is expressed in constant 2002 dollars, using the Producer Price Index for farm products.
[2] Distributions are based on the number of farms in 2002 that also existed in previous years. For example, the first bar shows the distribution of the 489 farms in 2002 that also existed in 1978.
Source: USDA, Economic Research Service, compiled from the 2002 Census of Agriculture Longitudinal File.

Figure 3. Farms with sales of $5 million or more in 2002, by constant-dollar sales class,[1] 1978-1997
Most $5-million farms had sales of at least $1 million in earlier years

SHIFT TO ARMS DATA

The rest of this report uses ARMS data rather than census data, since ARMS provides more recent and more detailed information. The census and ARMS provided similar estimates of the number of million-dollar farms. The ARMS count of million-dollar farms in 2002 was close to the census count, especially for farms with gross sales between $1 million and $4,999,999 (table 3). The census has traditionally done a better job covering the largest farms (Banker and MacDonald, 2005), because of intensive efforts to get responses from large or unique operations (USDA, NASS, 2004). By 2006, the ARMS count of million-dollar farms was 35,100. However, the difference between the 2002 and 2006 ARMS estimates was not statistically significant.[5]

Share of Farms, Gross Farm Sales, and Government Payments

Like the 2002 Census of Agriculture, ARMS data show million-dollar farms making up a disproportionately large share of gross farm sales, given their small share of farms. Million-dollar farms made up 2 percent of all U.S. farms in 2006 and held 13 percent of farm assets (including land), but reported 48 percent of farm product sales (figure 4).

Government Payments

Million-dollar farms receive only 16 percent of U.S. Government payments to farmers (table 4), which is small compared with their 48-percent share of sales. Most Government payments are commodity-related or targeted at current or past production of specific commodities, largely feed and food grains, cotton, and oilseeds (see box, "Farm Program Payments"). Receipt of commodity-related payments historically has been proportional to harvested acres of program crops. Million-dollar farms, particularly those with sales of $5 million or more, harvest a small share of the acres supporting these crops. Thirty-five percent of million-dollar farms (including 53 percent of $5-million farms) receive no Government payments at all compared with the 21- or 27-percent share for other large-farm sales classes.

Table 3. Number of million-dollar farms, by sales class and data source, 2002

Sales class	2002 census	2002 ARMS	2002 ARMS/ 2002 census
	Number		Percent
Million-dollar farms	28,673	27,202	94.9
$1,000,000 to 4,999,999	25,335	*25,211	99.5
$5,000,000 or more	3,338	1,991	59.6

* = standard error is between 25 and 50 percent of the estimate.
Source: 2002 Census of Agriculture and 2002 Agricultural Resource Management Survey, Phase III.

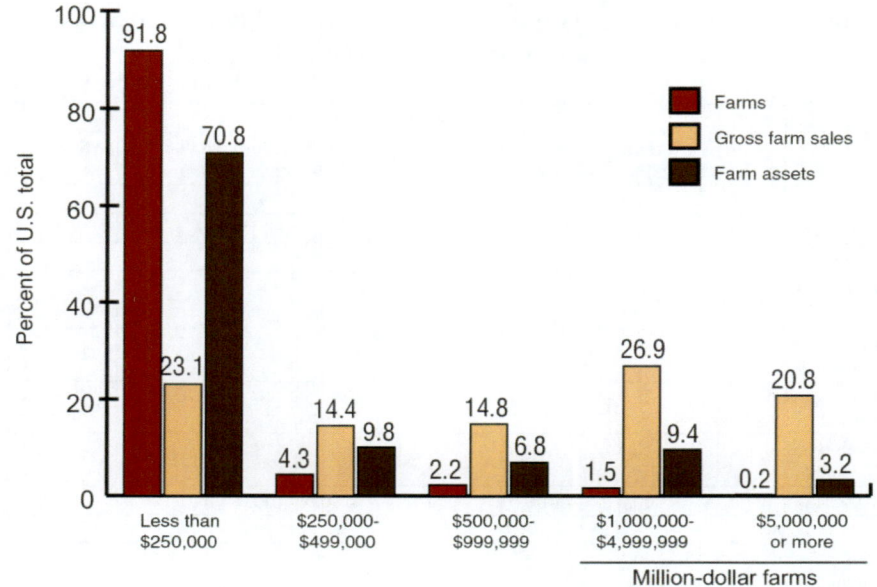

Note: Items may not add to totals due to rounding.
Source: USDA, Economic Research Service, 2006 Agricultural Resource Management Survey, Phase III.

Figure 4. Distribution of farms, gross farm sales, and farm assets, by sales class, 2006
Million-dollar farms account for 2 percent of farms, but 48 percent of sales

Individual Commodities

Million-dollar farms also accounted for a 48-percent share of the value of production, a measure similar to sales that excludes the effects of inventory change on sales.[6] As shown in figure 5, million-dollar farms account for even more of the value of production for particular commodities: high-value crops (72 percent), dairy products (59 percent), hogs (58 percent), poultry (55 percent), and beef (52 percent). The larger five-million-dollar farms alone account for 41 percent of the sales of high-value crops, 35 percent of the sales of beef cattle, and 27 percent of milk production. According to the 2002 Census of Agriculture, a large share of million-dollar beef farms are feedlots: 44 percent of farms with sales between $1 million and $4,999,999 and 83 percent for farms with sales of $5 million or more.

Table 4. Distribution of Government payments and harvested acres of program crops, by sales class, 2006

Item	Less than $250,000	$250,000-$499,999	$500,000-$999,999	$1,000,000 or more			All farms
				All $1,000,000-$4,999,999		$5,000,000 or more	
Number							
Total farms	1,912,457	90,239	45,857	35,121	31,145	3,976	2,083,674
Percent of farms in sales class							
Payments received:							
None	60.1	21.0	26.8	35.4	33.2	52.5	57.3
Only conservation[1]	9.5	1.3	2.0	3.1	3.2	1.9	8.9
Only commodity-related[1]	23.9	53.4	47.2	43.5	44.4	36.4	26.0
Both types of payments	6.5	24.3	23.9	18.1	19.2	9.2	7.9
All farms	100.0	100.0	100.0	100.0	100.0	100.0	100.0
Percent of U.S. total							
Government payments	46.6	20.5	16.7	16.2	14.3	1.9	100.0
Conservation[1]	78.5	9.5	6.2	5.7	4.8	0.9	100.0
Commodity-related[1]	38.8	23.2	19.2	18.8	16.6	2.2	100.0
Harvested acres of program crops[2]	36.2	26.6	20.4	16.8	15.5	1.3	100.0

Note: Items may not add to totals due to rounding.

[1] For definitions of conservation program payments and commodity-related payments, see box below.

[2] Corn, cotton, peanuts, rice, sorghum, soybeans, tobacco, barley, oats, wheat, canola, and other oilseeds.

Source: USDA, Economic Research Service, 2006 Agricultural Resource Management Survey, Phase III.

The large share of dairy, beef, hog, and poultry production by million-dollar farms refl ects the movement of livestock production from an open environment to climate-controlled buildings, which makes production less dependent on the weather. Other technologies—disease control, handling, transportation, and nutrition—have increased the number of production cycles per year. These technological advancements helped standardize production, making it easier for farms to operate on a large scale (Allen and Lueck, 1998).

High-value crops—other than some horticultural specialties—are generally produced outdoors, like other crops. Other characteristics of these crops, however, make their production more routine, encouraging large-scale farming (Allen and Lueck, 1998). High-value crops are often irrigated, which reduces the variability of harvest. These crops may require a large amount of labor relative to other crops, but the labor is applied in a restricted area, which makes it easier to supervise. In areas like California, several plantings and harvests of vegetables may occur in a year, which means labor can be used on a more constant basis.

FARM PROGRAM PAYMENTS

The payments covered by the 2006 Agricultural Resource Management Survey (ARMS) can be sorted into two major categories.

1. Commodity-related: Direct payments, countercyclical payments, loan deficiency payments, marketing loan gains, net value of commodity certificates, milk income loss contract payments, agricultural disaster payments, and any other State, Federal, and local payments.
2. Conservation: Payments from the Conservation Reserve Program (CRP), Conservation Reserve Enhancement Program (CREP), Wetlands Reserve Program (WRP), Environmental Quality Incentives Program (EQIP), and Conservation Security Program (CSP).

Since ARMS contacts only farm operators, the survey excludes farm program payments made to people who do not farm, mainly nonoperator landlords.

SPECIALIZATION, FARM SIZE, AND TENURE

Due to the large share of specific commodities sold by million-dollar farms, they obviously specialize in different commodities than other large farms. Million-dollar farms specialize less in cash grains, but more in high-value crops and hogs, than farms in either of the other large-farm sales classes (table 5). Other major specializations for million-dollar farms include beef, dairy, and poultry. Million-dollar farms also account for a relatively large percentage of the farms in some specializations (figure 6). Six percent of farms specializing in high-value crops sell at least $1 million of products, as do 8 percent of dairy farms, 13 percent of cotton farms, 15 percent of hog farms, and 17 percent of poultry farms.

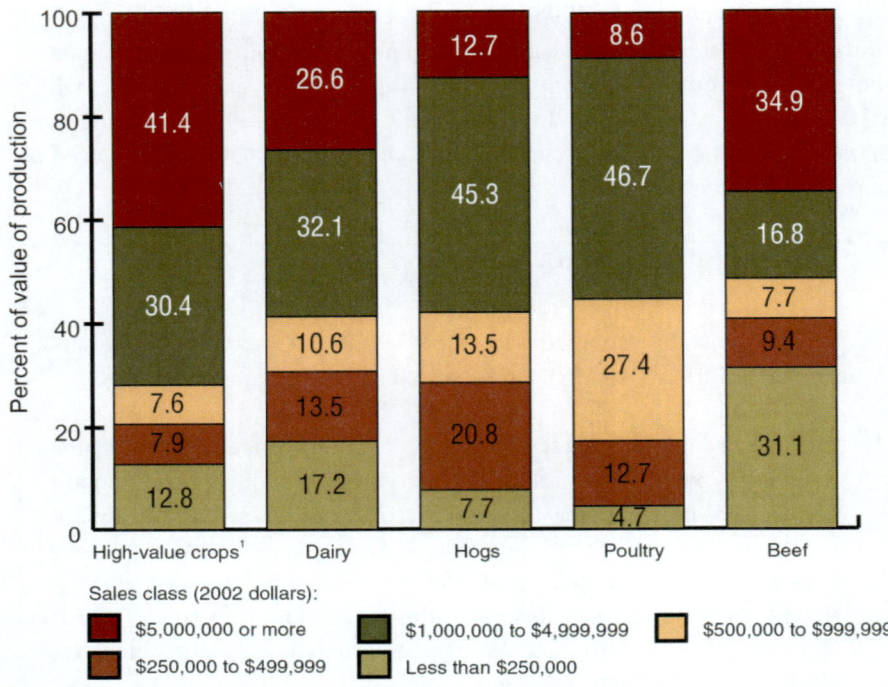

Note: Items may not add to totals due to rounding.
[1] Vegetables, fruits and tree nuts, and nursery and greenhouse products.
Source: USDA, Economic Research Service, 2006 Agricultural Resource Management Survey, Phase III.

Figure 5. Distribution of the value of production for selected commodities, 2006
Million-dollar farms sell most of several commodities

Farms with more than $5 million in sales have concentrated on three specializations—high-value crops, beef, and dairy—which account for 82 percent of the farms in this sales class (table 5). The prevalence of these specializations among $5-million farms suggests major economies of scale in the production of high-value crops, finished beef cattle, and milk even when sales pass $5 million.

Farm Size

As one might have anticipated, average acreage operated increases with sales volume. Average farm size increases from 281 acres for farms with sales less than $250,000 to 3,400 acres for all million-dollar farms (table 6). Average acreage increases as sales increase from the $1,000,000-$4,999,999 level to $5 million or more, but this increase is not statistically significant.

Table 5. Farm specialization, by sales class, 2006

Item	Less than $250,000	$250,000-$499,999	$500,000-$999,999	$1,000,000 or more All	$1,000,000-$4,999,999	$5,000,000 or more	All farms
			Number				
Total farms	1,912,457	90,239	45,857	35,121	31,145	3,976	2,083,674
			Percent				
Commodity specialization:[1]							
Cash grain[2]	11.0	42.0	34.8	13.7	15.3	d	12.9
Cotton	0.2	3.6	5.8	4.2	4.6	d	0.6
Other field crops[3]	25.8	5.0	8.3	6.2	6.8	d	24.2
High-value crops[4]	5.4	9.2	9.0	20.2	18.3	34.9	5.9
Beef	35.7	14.1	10.8	15.0	13.7	25.1	33.9
Hogs	0.6	3.9	4.6	8.2	8.6	d	0.9
Dairy	2.0	12.6	9.5	13.8	12.9	21.5	2.8
Poultry	0.8	7.3	15.4	17.0	18.5	5.3	1.7
Other livestock[5]	18.5	2.3	1.7	1.6	1.2	d	17.1
All farms	**100.0**	**100.0**	**100.0**	**100.0**	**100.0**	**100.0**	**100.0**

Note: Items may not add to totals due to rounding.
d = data suppressed due to insufficient observations.
[1] Commodity that accounts for at least half of the farm's value of production.
[2] Includes wheat, corn, soybeans, grain sorghum, rice, and general cash grains, where no single cash grain accounts for the majority of production.
[3] Tobacco, peanuts, sugar beets, sugar cane, corn for silage, sorghum for silage, hay, canola, and general crops, where no single crop accounts for the majority of production. Also includes farms with all cropland in the Conservation Reserve Program (CRP), Conservation Reserve Enhancement Program (CREP), and Wetlands Reserve Program (WRP).
[4] Vegetables, fruits and tree nuts, and nursery and greenhouse products.
[5] Includes sheep, lambs, wool, goats, goats' milk, mohair, horses, ponies, mules, donkeys, bees, honey, aquaculture, mink, rabbits, other fur-bearing animals, bison, deer, elk, llamas, etc. Also includes farms where no single livestock species accounts for the majority of production.
Source: USDA, Economic Research Service, 2006 Agricultural Resource Management Survey, Phase III.

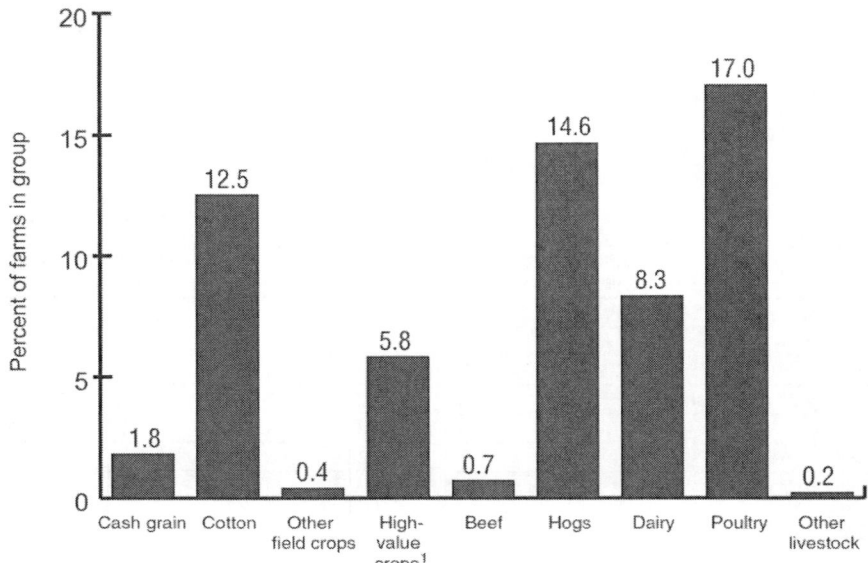

[1] Vegetables, fruits and tree nuts, and nursery and greenhouse products.
Source: USDA, Economic Research Service, 2006 Agricultural Resource Management Survey, Phase III.

Figure 6. Share of farms with sales of $1 million or more, by specialization, 2006
Million-dollar farms make up a large percentage of poultry, hog, cotton, dairy, and high-value crop farms

Average acreage operated is not the best indicator of the size of a typical farm in a group, because a few high-acreage farms in a particular group can raise the average well above the acreage operated on most farms. Median acreage operated—the midpoint of the distribution of farms by acres operated—is a better indicator of farm size. In the case of million-dollar farms, average acres operated greatly exceeds median acres operated because of a few extensive cattle ranches.

A different pattern between acreage and sales class emerges if medians are used. Median acres operated increases with sales until the $1-million level. The median for million-dollar farms is approximately 200 acres less than the corresponding estimate for farms with sales between $500,000 and $999,999. A larger share of million-dollar farms specialize in relatively low-acreage commodities (high-value crops, hogs, and dairy) than farms with sales just under $1 million, whose main specialization is cash grain, a high-acreage commodity.

Table 6. Farm acreage and tenure, by sales class, 2006

Item	Less than $25,000	$250,000-$499,999	$500,000-$999,999	$1,000,000 or more All	$1,000,000-$4,999,999	$5,000,000 or more	All farms
				Number			
Total farms	1,912,457	90,239	45,857	35,121	31,145	3,976	2,083,674
				Percent of U.S. total			
Acres of farmland:							
Owned	68.7	11.4	7.9	12.1	10.7	1.4	100.0
Operated	59.6	16.0	11.0	13.4	11.7	1.6	100.0
				Acres per farm			
Average (mean) acres operated	281	1,602	2,168	3,430	3,398	3,682	432
Owned	203	712	968	1,946	1,933	2,052	271
Rented in	106	915	1,230	1,532	1,505	*1,743	190
Rented out	28	*25	30	48	39	*114	29
Median acres operated[1]	84	825	1,258	1,045	1,064	983	100
				Percent of group			
Tenure:							
Full owner	66.3	18.4	21.4	30.1	28.6	42.0	62.7
Part owner	27.8	70.5	65.8	56.3	58.6	38.2	31.0
Tenant[2]	5.8	11.1	12.8	13.6	12.8	19.7	6.3
All farms	100.0	100.0	100.0	100.0	100.0	100.0	100.0

Note: Items may not add to totals due to rounding.
* = standard error is between 25 and 50 percent of the estimate.
[1] Midpoint of the distribution of farms by acres operated. Half the farms in a group operate more acres than the median, while the other half operate fewer acres than the median.
[2] Farms that rent all the land they operate. Also includes farms owning less than 1 percent of the land they operate.
Source: USDA, Economic Research Service, 2006 Agricultural Resource Management Survey, Phase III.

Tenure

Renting is commonly used to control land without the debt and commitment of capital associated with land ownership. Full ownership of the land farmed is common among small farms but is less common among large farms, where a greater share of farms rent, either as part owners or tenants. Depending on the sales class, 20 to 30 percent of farms with sales greater than $250,000—including million-dollar farms considered as a whole—own all the land they farm.

Farms with sales greater than $5 million, however, are an exception: 42 percent are full owners. These full-owners farm a median of 460 acres or half the 980-acre median for all $5-million farms. Eighty percent of these $5-million full owners specialize in beef, diary, or high-value crops. Feedlots and dairy farms do not require large acreages if most of the feed is bought rather than grown. As pointed out earlier in the box, "A Million Dollars in Sales," specialty crops can generate high revenue on a small acreage.

BUSINESS ORGANIZATION

Most U.S. farms (92 percent) are sole proprietorships, but the share of farms organized as such declines with sales (table 7). For farms with $1 million or more in sales, only 45 percent of farms are sole proprietorships. Million- dollar farms are more commonly organized as partnerships or corporations than are smaller farms, and these forms of organization account for 64 percent of gross sales from million-dollar farms. C-corporations and S-corporations are most prevalent among farms with sales of $5 million or more. Fifty-one percent of $5-million farms are incorporated compared with only 31 percent of the smaller million-dollar farm.

U.S. farms are seldom part of a larger firm, such as a company that processes farm products. Even among smaller million-dollar farms—those with sales less than $5 million—only 3 percent report that they are a subsidiary of another company. Ten percent of $5-million farms, however, are part of larger companies.[7] About 82 percent of these $5-million subsidiaries specialize in either high-value crops or beef.

Limited liability companies (LLCs) are a relatively new form of organization allowed under State law (U.S. Department of the Treasury, Internal Revenue Service, 2008). LLCs provide business owners with limited liability for debts and actions of the business, management flexibility, and pass-through taxation. LLCs are not common among lower-sales farms and do not exceed 10 percent of farms until sales reach $1 million. They are most common among farms with sales of $5 million or more, 27 percent of which are LLCs.

Table 7. Business Organization of Farms, by sales class, 2006

Item				$1,000,000 or more			All farms
	Less than $250,000	$250,000-$499,999	$500,000-$999,999	All	$1,000,000-$4,999,999	$5,000,000 or more	
				Number			
Total farms	1,912,457	90,239	45,857	35,121	31,145	3,976	2,083,674
				Percent of group			
Farms by business organization:							
Sole proprietorship[1]	93.6	76.4	67.9	45.4	48.3	22.6	91.5
Legal partner-ship[2]	3.2	11.6	13.0	22.2	22.1	22.9	4.1
C-corporation[3]	0.7	6.0	9.2	15.1	14.3	21.4	1.3
S-corporation[3]	1.3	5.3	8.9	15.9	14.2	29.4	1.9
Other[4]	1.2	0.8	1.1	1.4	1.0	3.8	1.2
All farms	100.0	100.0	100.0	100.0	100.0	100.0	100.0
Gross farm sales by business organization:							
Sole proprietorship[1]	86.9	75.6	67.1	33.8	44.5	20.0	57.0
Legal partner-ship[2]	6.2	11.7	13.5	24.7	23.5	26.1	16.9
C-corporation[3]	2.1	6.4	9.4	18.7	14.7	23.9	11.7
S-corporation[3]	3.1	5.5	8.9	20.7	15.9	26.9	12.7
Other[4]	1.7	0.7	1.1	2.1	1.3	3.1	1.6
All sales	100.0	100.0	100.0	100.0	100.0	100.0	100.0
Farm is part of larger firm or corporation:[5]							
Share of farms	1.4	d	d	3.8	3.1	*10.4	1.4
Share of gross farm sales	0.9	d	d	*8.3	3.4	*15.6	4.3
Limited liability company:[6]							
Share of farms	2.1	5.8	7.7	14.5	13.0	26.9	2.6
Share of gross farm sales	3.6	5.7	7.7	21.6	14.4	31.0	13.1
Family farm:[7]							
Share of farms	97.5	95.5	92.2	84.3	86.8	64.3	97.1
Share of gross farm sales	94.9	95.4	92.1	72.9	85.0	57.1	84.0

Note: Items may not add to totals due to rounding.

d = data suppressed due to insufficient observations.
* = standard error is between 25 and 50 percent of the estimate.
[1] Includes informal partnerships, such as those between spouses.
[2] Includes only partnerships registered under State law.
[3] A C-corporation is legally separate and distinct from its owners, shareholders, or stockholders. The corporation is formed by filing articles of incorporation. An S-corporation—or small business corporation—provides the benefits of incorporation while being taxed like a partnership or sole proprietorship.
[4] Estates, trusts, cooperatives, and grazing associations.
[5] Excludes contractual arrangements with totally separate firms. From version 1 of the 2006 ARMS.
[6] Limited liability companies (LLCs) are also reported in the more traditional categories above (proprietorships, partnerships, etc.), which LLCs use when paying taxes.
[7] Any farm where the majority of the business is owned by the operator and individuals related to the operator.
Source: USDA, Economic Research Service, 2006 Agricultural Resource Management Survey, Phase III.

Ninety-seven percent of all farms are "family farms," where the majority of the business is owned by the operator and the operator's relatives (see box, "What Is a Family Farm?"). Proportionally fewer million-dollar farms are family operations, but 84 percent are still family-operated and these family farms account for 73 percent of the gross sales from million-dollar farms. Family farms make up a smaller share of all farms and production once sales pass $5 million. Nevertheless, most $5-million farms are still family farms.

Nearly half of the nonfamily million-dollar farms are also nonfamily corporations (figure 7). These nonfamily corporations, however, are not large, publicly held companies; 89 percent had no more than 10 stockholders. Even nonfamily farm corporations with sales of at least $5 million usually had fewer than 10 stockholders.

WHAT IS A FAMILY FARM?

There is no hard and fast definition of "family farm." The ideal definition would allow for changes in the way operators structure their farm businesses as they respond to changes in technology, the marketplace, and policies, but still capture the general concept of a family farm in which a family maintains majority control and ownership.

> The definition of family farm used by the Economic Research Service (ERS) has changed over time. The current definition, as used in this report, includes any farm where the majority of the business is owned by the operator—or the principal operator on multi-operator farms—and individuals related to the operator by blood or marriage, including relatives who do not live in the operator's household. In 2006, 97 percent of farms in the Agricultural Resource Management Survey (ARMS) were classified as family farms under this definition.
>
> Prior to the adoption of the current definition, all farms were family farms, unless they were organized as cooperatives or nonfamily corporations, held in estates or trusts, or operated by a hired manager. ARMS typically classified 98 percent of farms as family farms using this definition.

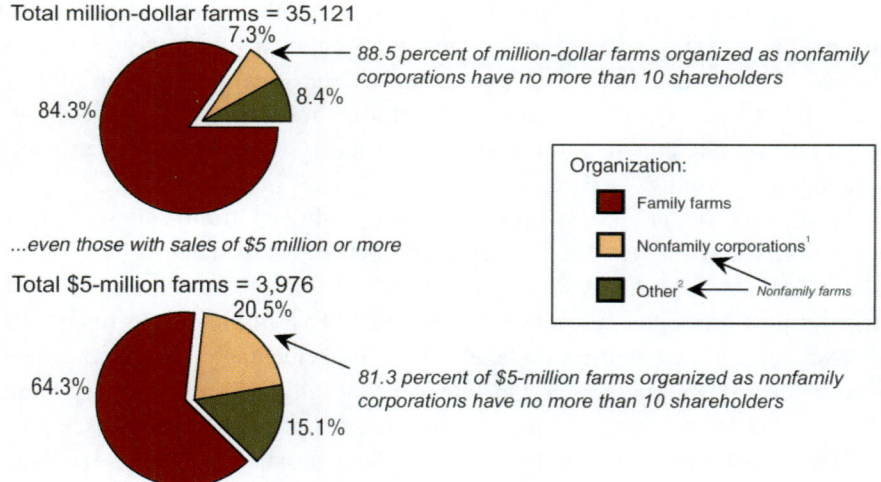

Note: Items may not add to totals due to rounding.

[1] Corporations where the operator and their relatives do not have a majority ownership interest.

[2] Estates, trusts, cooperatives, grazing associations, and any unincorporated farm businesses where the operators and their families do not hold a majority ownership interest.

Source: USDA, Economic Research Service, 2006 Agricultural Resource Management Survey, Phase III. (Number of shareholders from version 1.)

Figure 7. Organization of million-dollar farms, 2006

Most million-dollar farms are family farms...

OPERATOR CHARACTERISTICS

Every farm has at least one operator, or a farmer who makes day-to-day decisions about the farm business. Some farms—particularly larger ones—have more than one operator who makes decisions. In such cases, one operator is designated as the principal operator, the one most responsible for running the farm, and the others are secondary operators. The count of principal operators also includes sole operators on single-operator farms.

Principal Operators

Principal operators of million-dollar farms are similar to their counterparts on other large farms. The average age of operators in the large-farm sales classes falls between 51 and 53 years, with no statistically significant differences among sales classes (table 8). Similarly, most operators of million-dollar farms—like the operators of other large farms—report that their primary occupation is farming.

The share of large-farm operators who graduated from college ranges between 28 and 33 percent for each sales class. One major difference in educational attainment exists between operators of $5-million farms and other large farms. Forty-eight percent of the operators of $5-million farms reported a high-school diploma (but no college), 10 or 11 percentage points more than operators of other large farms. Operators of $5-million farms may rely more on experience over formal education than operators of other large farms.

The largest differences in demographic characteristics occur between large and small farms, not among the various large-farm sales classes. For example, 29 percent of small-farm operators are at least 65 years old, compared with 10 to 16 percent of large-farm operators; and 39 percent of small-farm operators report farming as their primary occupation, compared with nearly all large-farm operators.

Secondary Operators

In addition to principal farm operators, secondary operators work on approximately 950,500 multiple-operator farms (table 9). The number of operators per farm increases with farm size, because commercial-size farms often require more management and labor than one individual can provide. The number of operators per farm reaches 2.1 operators—on average—for million-dollar farms as a whole and peaks at 2.6 operators for farms with sales greater than $5 million. Multiple-operator farms account for a 66-percent share of million-dollar farms.

About 6 percent of all farms (and 14 percent of multiple-operator farms) are multiple-generation farms, with at least 20 years' age difference between the oldest and youngest operators. Multiple-generation farms make up a larger share of million-dollar farms (23 percent) than any other sales class (Figure 8), probably because million-dollar farms have a large enough business to support the financial needs of more than one generation.

Because farms are generally family businesses, one would expect family members to serve as secondary operators. In fact, 75 percent of all secondary operators on small farms are spouses (table 9). Although the share of large farms where the spouse is an operator is fairly constant—roughly 30 percent of farms until sales reach $5 million—the spousal share of secondary operators declines as sales increase and secondary operators other than spouses are added. For all million-dollar farms, only 26 percent of secondary operators are spouses.

Spouses work as operators on only 11 percent of $5-million farms and make up a 7-percent share of secondary operators in that sales class. Five-milliondollar farms are less likely to be family farms than other farms—as discussed earlier—which means that the farms are not closely held by the operators and their households. As a result, there may be less financial incentive for their household members to participate in the farm business. In addition, farms with sales of $5 million—family or nonfamily—may require more time from secondary operators than spouses can provide.

FARM AND HOUSEHOLD FINANCES

Farm profits are strongly associated with farm size (figure 9). The three sales classes below $25,000 operate at a large percentage loss. The profit margin remains negative, although to a lesser degree, until sales reach $175,000. The average profit margin then increases to 20 percent for million-dollar farms with sales less than $5 million and peaks at 26 percent for farms with sales of $5 million or more. The same general pattern—operating profit margin increasing with sales—applies regardless of specialization.

Table 8. Age, education, and occupation of principal operators, by sales class, 2006

Item	Less than $250,000	$250,000-$499,999	$500,000-$999,999	$1,000,000 or more			All farms
				All	$1,000,000-$4,999,999	$5,000,000 or more	
				Number			
Total principal operators	1,912,457	90,239	45,857	35,121	31,145	3,976	2,083,674
				Years			
Average age	57	53	52	52	53	51	57
				Percent of group			
Age:							
Younger than 35 years	4.2	8.1	6.3	6.1	5.6	*10.1	4.5
35 to 44 years	11.1	16.5	17.0	17.1	17.4	15.2	11.6
45 to 54 years	26.5	31.3	36.5	35.5	34.8	40.4	27.0
55 to 64 years	29.0	28.1	25.7	27.3	27.7	24.7	28.9
65 years or older	29.2	16.0	14.6	13.9	14.5	*9.7	28.1
All principal operators	100.0	100.0	100.0	100.0	100.0	100.0	100.0
Operator is retired	21.5	3.0	3.0	1.9	2.1	0.8	20.0
Occupation:							
Farm or ranch work	39.0	91.5	95.7	96.4	96.2	98.4	43.5
Work other than farming	48.6	7.9	4.0	3.1	d	d	45.1
Not in the paid workforce	12.4	0.5	0.3	0.5	d	d	11.4
All principal operators	100.0	100.0	100.0	100.0	100.0	100.0	100.0
Education:							
Less than high school diploma	10.5	4.8	4.5	4.6	4.9	1.5	10.0
High school diploma	41.9	37.4	38.2	39.1	37.9	48.0	41.6

Some college	22.7	27.8	28.8	25.4	26.5	17.1	23.1
College graduate and beyond	25.0	30.1	28.4	31.0	30.7	33.3	25.4
All principal operators	100.0	100.0	100.0	100.0	100.0	100.0	100.0

Note: Items may not add to totals due to rounding.
d = data suppressed due to insufficient observations.
* = standard error is between 25 and 50 percent of the estimate.
Source: USDA, Economic Research Service, 2006 Agricultural Resource Management Survey, Phase III.

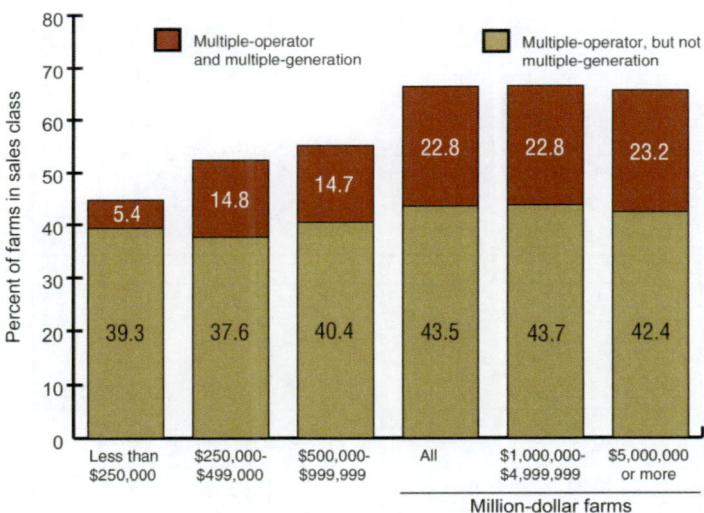

Note: Multiple-operator farms are defined as farms with more than one operator. Multiple-generation farms are multiple-operator farms with a difference of at least 20 years between the ages of the youngest and oldest operators.
Source: USDA, Economic Research Service, 2006 Agricultural Resource Management Survey, Phase III.

Figure 8. Multiple-operator and multiple-generation farms, by sales class, 2006
Multiple-generation farms are most common among million-dollar farms

Table 9. Multiple-Operator Farms, by Sales Class, 2006

Item	Less than $250,000	$250,000-$499,999	$500,000-$999,999	$1,000,000 or more All	$1,000,000-$4,999,999	$5,000,000 or more	All farms
			Number				
Total operators[1]	2,838,076	147,857	78,400	72,932	62,426	10,506	3,137,264
Principal operators[2]	1,912,457	90,239	45,857	35,121	31,145	3,976	2,083,674
Secondary operators	925,619	57,618	32,543	37,811	31,281	6,530	1,053,590
Spouses	698,139	28,319	14,533	9,639	9,195	443	750,630
Other	227,479	29,299	18,010	28,172	22,086	6,086	302,960
			Percent of farms				
Spouse is an operator[3]	36.5	31.4	31.7	27.4	29.5	11.2	36.0
			Percent of secondary operators				
Spousal share of secondary operators	75.4	49.1	44.7	25.5	29.4	*6.8	71.2
			Number				
Operators (principal and secondary) per farm	1.5	1.6	1.7	2.1	2.0	2.6	1.5
			Percent of farms				
Farms by number of operators:							
One	55.3	47.6	44.9	33.6	33.5	34.5	54.4
Two	41.6	42.7	43.5	43.6	44.8	33.6	41.7
Three	2.5	8.2	8.5	15.0	14.8	16.1	3.1
Four or more	0.5	1.5	3.1	7.8	6.8	15.8	0.8
All farms	100.0	100.0	100.0	100.0	100.0	100.0	100.0
			Number				
Multiple-operator farms[4]	854,704	47,260	25,264	23,304	20,698	2,606	950,533
			Percent				
Multiple-operator farms' share of:							
All farms	44.7	52.4	55.1	66.4	66.5	65.5	45.6
Gross farm sales	46.1	52.7	55.4	64.9	66.6	62.8	57.4

Note: Items may not add to totals due to rounding.

* = standard error is between 25 and 50 percent of the estimate.

[1] The Agricultural Resource Management Survey counts all operators–principal and secondary–and asks for detailed information on up to three operators.

[2] The number of principal operators equals the number of farms. Each farm has one principal operator.
[3] Calculated for farms with or without a spouse present.
[4] Mulitiple-operator farms report more than one operator.
Source: USDA, Economic Research Service, 2006 Agricultural Resource Management Survey, Phase III.

Standard Financial Performance Measures

A pattern similar to that for the operating profit margin also appears for other profitability measures, even when fewer sales classes are used (table 10). The rates of return on assets and equity are negative for farms with sales less than $250,000, but beyond that sales class they are positive, increase with sales, and are highest for million-dollar farms, particularly those with sales of $5 million or more.

In some respects, however, million-dollar farms are similar to other large farms. The share of farms with positive net farm income is fairly constant among large farms, just under 80 percent, regardless of sales class. The operating expense ratio varies in a fairly narrow range, from 73 to 79 percent once sales exceed $250,000. The situation is similar for the debt/asset ratio, which ranges from 11 to 15 percent—increasing with sales—for farms in the three sales classes between $250,000 and $4,999,999.

The debt/asset ratio, however, is higher for $5-million farms (25 percent) than for other large farms. The high debt/asset ratio for $5-million farms is also reflected in the large shares of farms classified as marginally solvent (25 percent) or vulnerable (8 percent). Farms in either of these categories—by definition—have debt/asset ratios greater than 40 percent.[8]

A high debt/asset ratio is not necessarily a problem, however, as long as the rate of return on assets exceeds the interest rate on the funds borrowed. On average, farms with sales greater than $5 million generate more net cash income per dollar of assets (or net worth) than other farms, and the larger gross cash income can be used to pay interest or reduce loan balances (figure 10). These farms are taking on more financial risk, but they also employ strategies to manage this risk. For example, about two-thirds of $5-million farms use marketing or production contracts. In addition, $5-million farms are more likely than smaller farms to be organized as corporations and LLCs, which means the operators' personal assets are not at risk.

Table 10. Selected financial performance measures, by sales class, 2006

Item	Less than $250,000	$250,000-$499,999	$500,000-$999,999	$1,000,000 or more			All farms
				All	$1,000,000-$4,999,999	$5,000,000 or more	
				Number			
Total farms	1,912,457	90,239	45,857	35,121	31,145	3,976	2,083,674
				Percent			
Profitability measures:							
Rate of return on assets[1]	-1.4	1.7	4.3	9.1	6.3	17.6	0.6
Rate of return on equity[2]	-1.9	0.9	3.8	9.5	6.0	21.0	0.0
Operating profit margin[3]	-23.1	8.6	16.5	22.6	20.0	26.2	4.5
				Dollars per farm			
Income measures:							
Gross cash farm income	31,271	343,069	647,748	2,511,632	1,644,037	9,307,770	100,149
Net farm income	7,801	70,042	150,006	601,502	360,779	2,487,159	23,633
				Percent			
Farms with positive net farm income	63.6	77.7	79.8	78.8	78.7	79.6	64.8
Financial efficiency measure:							
Operating expense ratio[4]	94.4	79.1	75.7	74.4	75.2	73.3	81.0
				Dollars per farm			
Balance sheet:							
Total assets	662,677	1,951,205	2,663,172	6,420,914	5,400,336	14,415,423	859,563
Total liabilities	36,767	220,368	365,605	1,095,585	783,490	3,540,324	69,803
Net worth	625,909	1,730,837	2,297,566	5,325,329	4,616,846	10,875,100	789,761
				Percent			
Solvency measure:							
Debt/asset ratio[5]	5.5	11.3	13.7	17.1	14.5	24.6	8.1
Solvency and income measure:							
Financial position:[6]							
Favorable	61.4	70.8	69.6	64.1	65.3	55.1	62.0
Marginal-income	33.4	17.8	16.2	14.2	14.4	*12.6	32.1

Item	$1,000,000 or more						
	Less than $250,000	$250,000-$499,999	$500,000-$999,999	All	$1,000,000-$4,999,999	$5,000,000 or more	All farms
Marginal-solvency	2.2	6.9	10.1	14.7	13.4	24.5	2.8
Vulnerable	2.9	4.5	4.1	7.0	6.9	7.9	3.1
All farms	100.0	100.0	100.0	100.0	100.0	100.0	100.0

Note: Items may not add to 100 due to rounding.
* = standard error is between 25 and 50 percent of the estimate.
[1] Return on assets = 100% X (net farm income + interest paid - charge for unpaid operators' labor and management) / total assets.
[2] Return on equity = 100% X (net farm income - charge for unpaid operators' labor and management) / net worth.
[3] Operating profit margin = 100% X (net farm income + interest paid - charge for unpaid operators' labor and management) / gross farm income.
[4] Operating expense ratio = 100% X total cash operating expenses / gross cash farm income.
[5] Debt/asset ratio = 100% X total liabilities/total assets.
[6] Financial performance classification based on farm income and debt/asset ratio:
- Favorable: positive net farm income and debt/asset ratio of no more than 40 percent.
- Marginal-income: negative net farm income and debt/asset ratio of no more than 40 percent.
- Marginal-solvency: positive net farm income and debt/asset ratio greater than 40 percent.
- Vulnerable: negative net farm income and debt/asset ratio greater than 40 percent.

Source: USDA, Economic Research Service, 2006 Agricultural Resource Management Survey, Phase III.

Farm Operator Household Income

The median income for households operating million-dollar farms was high in 2006: approximately $151,800 per household for those with farm sales between $1 million and $4,999,999 and $572,700 per household for those with sales of $5 million or more (figure 11).[9] In contrast, the median household income was just $54,800 for all U.S. farm households—as reported in the 2006 ARMS—and $48,200 for all U.S. households (DeNavas-Walt et al., 2007). The income of million-dollar households came largely from farm sources. Households operating the two groups of million-dollar farms each received a median off-farm income of roughly $25,000, which is on a par with other large-farm households, but less than half the $53,500 median for all U.S. farm households.

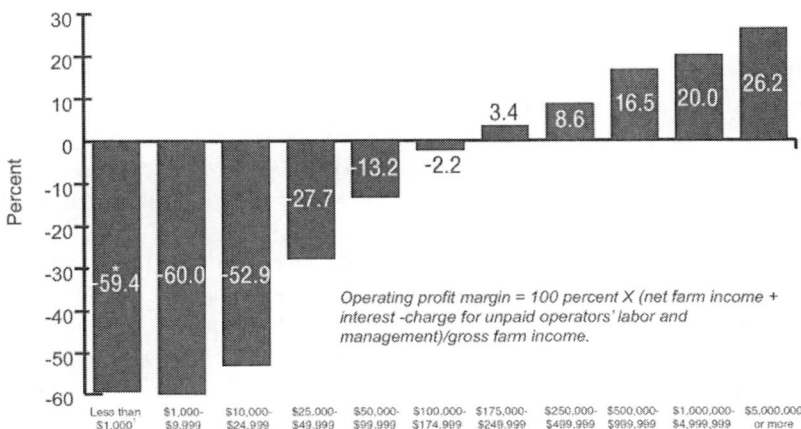

*= standard error is between 25 and 50 percent of the estimate.
[1] Point farms have sales of less than $1,000, but are still considered farms because they would be expected to normally sell at least $1,000 of agricultural products.
Source: USDA, Economic Research Service, 2006 Agricultural Resource Management Survey, Phase III.

Figure 9. Operating profit margin, by sales class, 2006
Operating profit margin increases with sales, once sales pass $10,000

Figure 11 explains how farm households selling less than $175,000 in sales can continue to operate, despite their negative average operating profit margins. Operators of these small farms do not completely exit farming because they have substantial off-farm income—particularly operators of farms with less than $100,000 in sales—and because they may be farming for reasons other than net income. Among these reasons are the potential for capital gains, losses from farming to write-off against other income for taxation purposes, and a rural lifestyle (Ahearn et al., 2004; Hoppe et al., 2005). Many small farms stay in business as long as the operator households have other sources of income and farm losses are not unduly and persistently large. The $1,000 sales cutoff in the farm definition means that many small farms are actually rural residences rather than farm businesses.

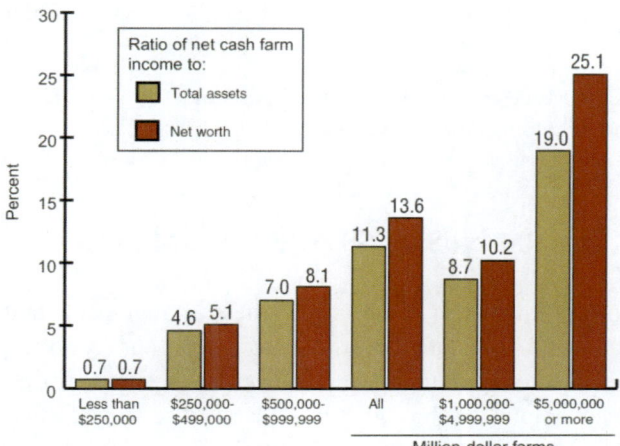

Note: Net cash income was adjusted by adding interest expense back in.
Source: USDA, Economic Research Service, 2006 Agricultural Resource Management Survey, Phase III.

Figure 10. Ratio of net cash farm income to assets and net worth, by sales class, 2006
The ratios are highest for $5-million farms

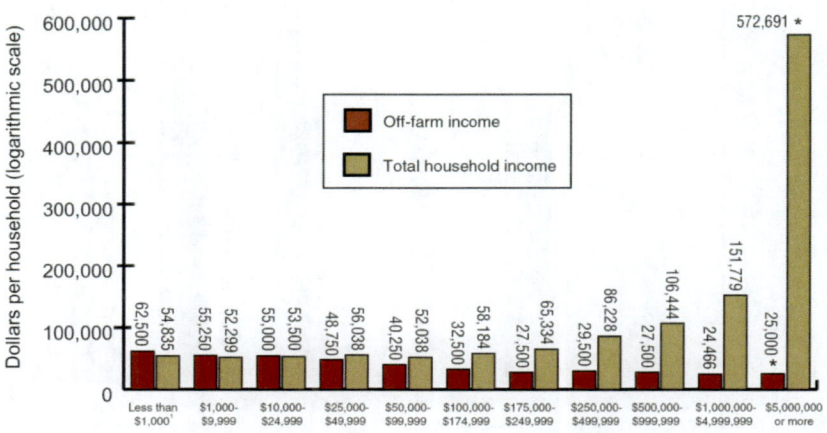

Note: A logarithmic scale is used because of the large range in total household income. Farm earnings are not shown, because negative values cannot be plotted on log charts, and farm earnings are negative for the first three sales classes.
*= standard error is between 25 and 50 percent of the estimate.
[1] Point farms have sales of less than $1,000, but are still considered farms because they would be expected to normally sell at least $1,000 of agricultural products.

Source: USDA, Economic Research Service, 2006 Agricultural Resource Management Survey, Phase III.

Figure 11. Median income of principal operator households, by sales class, 2006
Total operator household income increases with sales for large farms

FARM BUSINESS ARRANGEMENTS

Million-dollar farms use a variety of business arrangements that link them to other firms and individuals in order to access or control productive resources. The key to agricultural production is the control of assets, but control can be accomplished through renting land (discussed earlier) and other assets rather than buying them outright. Similarly, farms can use hired/contract labor or custom work rather than family labor. Farms may also link to other firms through marketing or production contracts to sell or remove the commodities they produce. Farm operations can also serve as contractors themselves.

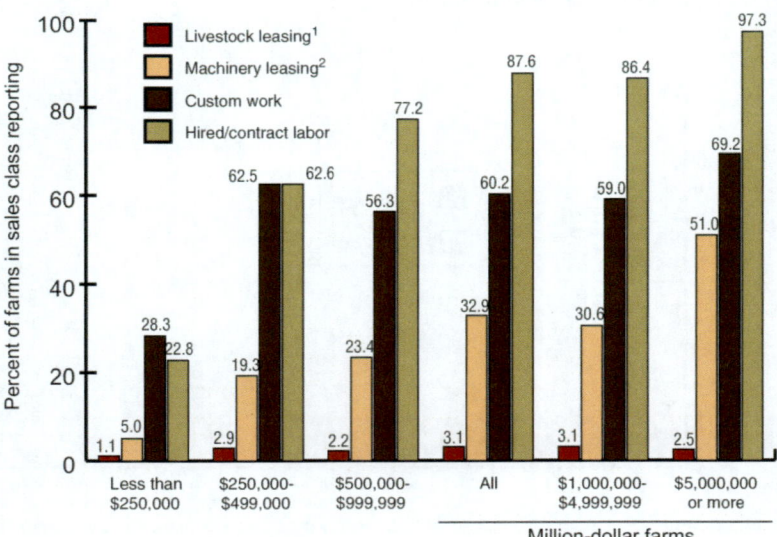

[1] Includes leasing bees for pollination.
[2] Renting or leasing tractors, vehicles, farm machinery and equipment, and storage structures.
Source: USDA, Economic Research Service, 2006 Agricultural Resource Management Survey, Phase III.

Figure 12. Selected methods of input procurement, by sales class, 2006
Machinery leasing, custom work, and hired/contract labor are most common among farms with sales greater than $5 million

Accessing Resources and Labor

Million-dollar farms often rely on machinery leasing, custom work, and hired/contract labor (figure 12). Thirty-one percent of smaller million-dollar farms rent machinery, and the rental rate increases to 51 percent for farms with sales of at least $5 million. Fifty-nine percent of farms with sales between $1 million and $4,999,999 use custom work, approximately the same share as smaller large farms, but less than the 69-percent share for $5-million farms.

Farms rent livestock infrequently, regardless of the level of their sales. Farms specializing in dairy or high-value crops rent livestock most often, but their overall rental rates are still low. About 3 percent of all dairy farms rent livestock, and this share does not vary much by sales class. A 4-percent share of farms specializing in high-value crops also rent livestock, namely bees for pollination. The share renting bees, however, is more for million-dollar farms and other large farms (around 10 percent) than the 3-percent share for small farms.

Hours of Labor

The use of hired/contract labor increases steeply with sales class, starting at 23 percent of farms with sales less than $250,000 and reaching 97 percent for $5-million farms. On average, million-dollar farms use about 36,500 hours of labor per farm per year or 18.2 "annual person equivalents," where an annual person equivalent is defined as one person working 40 hours per week for 50 weeks (table 11). Most of the labor on million-dollar farms is either hired (72 percent) or contracted (13 percent), with most of the balance provided by principal and secondary operators.[10]

Smaller farms use less labor—as one would expect—and the principal operator and spouse account for a larger share of labor hours. Farms in the $500,000 to $999,999 sales class use 4.4 annual person equivalents, and principal operators and spouses account for 39 percent of the labor used, compared with just 10 percent on million-dollar farms.

Table 11. Sources of farm labor, by sales class, 2006

Item			$1,000,000 or more				All farms
	Less than $250,000	$250,000-$499,999	$500,000-$999,999	All	$1,000,000-$4,999,999	$5,000,000 or more	
	Number						
Total farms	1,912,457	90,239	45,857	35,121	31,145	3,976	2,083,674
	Annual hours per farm						
Mean hours worked	2,231	6,410	8,865	36,494	23,724	136,526	3,135
	Percent of total hours						
Share of total hours worked by:							
Principal operator[1]	57.1	45.5	34.1	8.4	13.0	2.2	45.1
Spouse[1]	14.2	8.3	4.6	1.3	2.1	0.3	10.6
Other operators[1]	4.8	11.2	10.4	5.1	7.3	2.0	5.8
Unpaid workers	5.5	3.5	2.5	0.7	1.1	0.1	4.2
Hired labor	16.6	28.2	41.7	71.8	65.0	81.2	30.0
Contract labor	1.9	3.3	6.7	12.6	11.5	*14.2	4.4
All sources	100.0	100.0	100.0	100.0	100.0	100.0	100.0
	Annual person equivalents per farm						
Total annual person equivalents[2]	1.115	3.205	4.432	18.247	11.862	68.263	1.568
	Annual person equivalents[2] per $100,000 of gross sales						
Labor per $100,000 in gross sales	4.187	0.907	0.624	0.608	0.622	0.591	1.480
	Percent of farms						
Farms by annual person equivalents:[2]							
Less than 5	98.9	88.2	73.9	40.2	44.2	8.9	96.9
5 to 9.999	0.9	10.2	19.5	24.7	26.0	14.8	2.1
10 to 19.999	d	d	5.4	15.6	15.5	16.2	0.5
20 or more	d	d	1.2	19.5	14.3	60.1	0.5
All farms	100.0	100.0	100.0	100.0	100.0	100.0	100.0

Note: Items may not add to totals due to rounding.
d = data suppressed due to insufficient observations.
* = standard error is between 25 and 50 percent of the estimate.
[1] Includes paid and unpaid hours.
[2] One annual person equivalent equals 2,000 hours or 50 weeks per year times 40 hours per week.

Source: USDA, Economic Research Service, 2006 Agricultural Resource Management Survey, Phase III.

Farms with sales of $5 million or more use about 68 annual person equivalents of labor, nearly 6 times that used by farms with sales between $1 million and $4,999,999. In part, this is simply a reflection of the size of $5-million farms, since the labor necessary to produce $100,000 of sales is similar—about .6 annual person equivalents—for both sales classes.

Capital/Labor Ratio

Labor is used in conjunction with assets. The capital used per annual labor equivalent is lower for million-dollar farms than for smaller farms, due largely to a lower value for real estate (figure 13). The ratio is particularly low for farms with sales of $5 million or more ($212,000 per annual labor equivalent).

High-value crop farms account for 60 percent of labor used on million-dollar farms, and these farms dominate the labor statistics for the group. Because high-value crop farms tend to be labor intensive rather than capital intensive compared with other specializations—such as cash grain farms (table 12)—they pull down the capital/labor ratio for all million-dollar farms.

Contracting

Contracting can provide benefits to both producers and contractors (MacDonald and Banker, 2005). Farmers have a guaranteed outlet for their production with known compensation, while contractors get an assured supply of commodities with specified characteristics, delivered in a timely manner. ERS identifies two types of contracts in ARMS:

1. **Production contract:** A legal agreement between a farm operator (contractee) and another person or firm (contractor) to produce a specific type, quantity, and quality of agricultural commodity. The contractor usually owns the commodity being produced and the farm receives a service fee.[11]

2. **Marketing contract:** The contractor buys a known quantity and quality of a commodity from a farm for a negotiated price (or pricing mechanism). The farm owns the commodity while it is

being produced and receives a price that refl ects the value of the commodity.

Contracting is common among million-dollar farms and farms with sales just under $1 million. Sixty-three percent of million-dollar farms—in both sales classes—have production or marketing contracts, about the same share as farms with sales from $500,000 to $1 million, but more than the share for smaller farms (table 13). About half of the value of production on million-dollar farms is under contract, 7 percentage points higher than farms with sales just under $1 million. Livestock account for 70 percent of production under contract on million-dollar farms.

Although they make up only 10 percent of all farms with contracts, million- dollar farms account for 62 percent of the value of production under contract and a 40-percent share of production not under contract. Two commodities make up 61 percent of this noncontract production on million-dollar farms: high-value crops (37 percent) and beef (24 percent). Cash grain and beef account for most noncontract production—65 to 70 percent, depending on the sales class—on smaller farms.

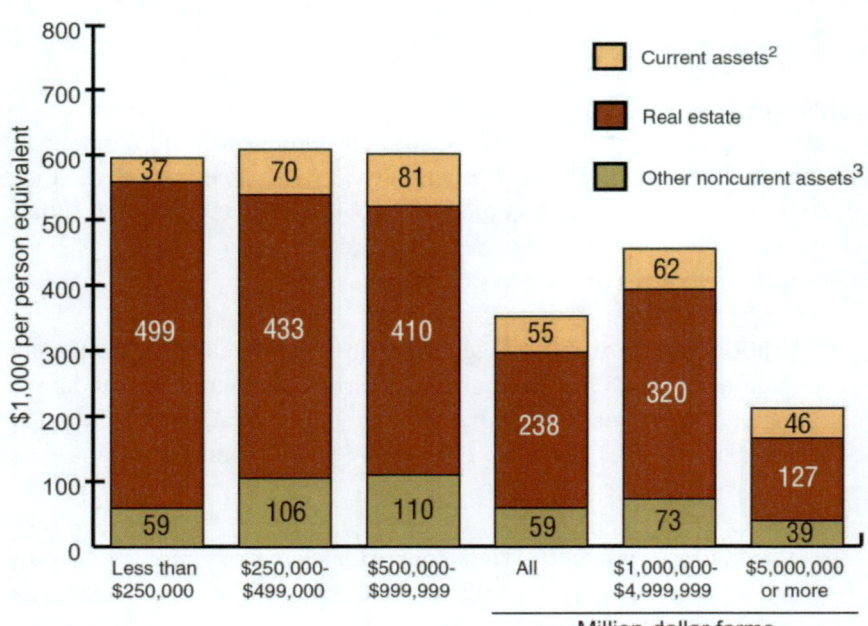

[1] Farm business assets divided by the number of annual person equivalents.
[2] Cash, assets that will be converted to cash within a year, and assets that will be used up within a year.
[3] Assets used in more than one year—other than real estate—such as machinery, equipment, and breeding stock.
Source: USDA, Economic Research Service, 2006 Agricultural Resource Management Survey, Phase III.

Figure 13. Capital/labor ratio,1 by type of asset and sales class, 2006
The ratio is lowest for million-dollar farms, especially farms with sales of $5 million or more

Table 12. Labor and capital on million-dollar farms specializing in cash grains or high-value crops, 2006

Item	Cash grain	High-value crops
	Number	
Number of farms	4,829	7,090
	Acres per farm	
Median acres owned	*700	160
	Annual person equivalents per $100,000 of sales	
Labor per $100,000 of sales	0.358	1.363
	Dollars per farm	
Assets per annual person equivalent:	836,333	158,643
Current asssets[1]	155,196	26,581
Real estate	509,417	117,718
Other noncurrent assets[2]	171,720	14,345

Note: Items may not add to totals due to rounding.
* = standard error is between 25 and 50 percent of the estimate.
[1] Mostly inventories.
[2] Mostly machinery and equipment.
Source: USDA, Economic Research Service, 2006 Agricultural Resource Management Survey, Phase III.

Farms can also serve as contractors. The 2006 ARMS questionnaire asked if any other any operations produced livestock—including poultry—under a contract arrangement for the farm being interviewed. Less than 1 percent of all U.S. farms reported acting as a contractor, but the percentage was higher as sales approached $1 million. Three percent of farms in the two sales classes between $500,000 and $4,999,999 contracted livestock production to other farms. The share increased to 12 percent for $5-million farms.

Table 13. Farms with contracts, by sales class, 2006

Item	Less than $250,000	$250,000-$499,999	$500,000-$999,999	$1,000,000 or more			All farms
				All	$1,000,000-$4,999,999	$5,000,000 or more	
Number							
Total farms	1,912,457	90,239	45,857	35,121	31,145	3,976	2,083,674
Percent of group							
Farms with contracts[1]	6.9	49.8	62.4	62.5	62.5	62.6	10.9
Value of production under contract[2]	17.5	32.5	42.9	50.0	47.0	53.7	39.0
Percent of U.S. total							
Farms with contracts[1]	58.1	19.7	12.6	9.6	8.5	1.1	100.0
Value of production under contract[2]	10.2	12.1	15.7	62.1	31.9	30.2	100.0
Value of production not under contract	30.9	16.0	13.4	39.8	23.1	16.7	100.0
Percent of group							
Farm acts as contractor[3]	d	d	3.3	4.2	*3.2	12.2	0.3
Percent of U.S. total							
Farm acts as contractor[3]	d	d	28.2	26.9	18.1	8.9	100.0

Note: Items may not add to totals due to rounding.
d = data suppressed due to insufficient observations.
[1] Farms reporting production under production contracts, marketing contracts, or both.
[2] Includes commodities under production or marketing contracts.
[3] Another operation grows livestock (including poultry) for the farm under a contract arrangement.

Source: USDA, Economic Research Service, 2006 Agricultural Resource Management Survey, Phase III.

CONCLUSIONS AND IMPLICATIONS

Three significant implications regarding million-dollar farms can be drawn from the information presented:

1. The shift in production to million-dollar farms is likely to continue. As long as the operating profit margin is proportional to sales class, million-dollar farms will have a competitive advantage. The shift in production may eventually slow, however, once million-dollar farms' shares of the commodities most amenable to large-scale production reach their upper limits.
2. There are still a sufficient number of million-dollar farms to prevent individual farms' domination agriculture or individual commodities. Concentration of production, however, may be a more significant conern when the owners of commodities—which include production contractors—are considered, rather than just the farms producing them.
3. Most million-dollar farms are family operations, although the operator and spouse supply only a small fraction of the labor. Direct ownership of million-dollar farms by nonfarm corporations is infrequent, but such corporations are frequently involved with million- dollar farms through contracting.

Continuing Shift to Million-Dollar Farms

The shift in production to larger farms has gone on for decades and is likely to continue. Million-dollar farms (and farms growing to that size) have a competitive advantage relative to smaller farms, refl ecting economies of size in farming.[12] These farms are able to take advantage of the forces that drive the structure of agriculture to a large-scale manufacturing model—identified by Gray and Boehlje (2007)—including technological change, economic/ financial innovations (such as outsourcing and multisite production), adapting general business management skills to farming, and coordinating farm production with suppliers and processors.

About three-quarters of current million-dollar farms specialize in five commodities: high-value crops, dairy, hogs, poultry, and beef (specifically feedlots). Million-dollar farms already account for half to three-fourths of the production of each of these commodities, and future shifts in the production of these commodities to million-dollar farms are likely. Eventually, however, some upper limit on the production of each commodity by million-dollar farms will be reached. The upper limit is currently unknown but will probably vary by commodity, and may approach 100 percent in some cases. At that point, any further increases in million-dollar farms' share of total agricultural production will come from producers specializing in other commodities. The positive relationship between the level of sales and operating profit margins applies for all specializations, so the competitive advantage for expanding farms also applies for specializations beyond the five listed.

Future shifts in production to million-dollar farms, however, may not be as dramatic as those seen from 1982 to 2002 once the upper limits are reached for the five commodities. Other commodities have characteristics that make them less amenable to large-scale production. Grain growers, for example, have only one production cycle per year and highly seasonal use of labor. Million-dollar grain farms existed in 2006, but they accounted for only 2 percent of all cash grain farms, versus 8 percent of dairy farms. In addition, 18 percent of the million-dollar dairies had sales of $5 million or more, but hardly any million-dollar grain farms sold that much. Shifts in production to million-dollar farms based on commodities less favorable to large-scale production may be more gradual, barring innovations in their production process.

Nevertheless, even if million-dollar farms' share of production slows or stabilizes, there still may be substantial upward shifts in production. As long as economies of scale exist for farms with sales of $5,000,000 or more, production can shift from the smaller million-dollar farms to those with sales greater than $5 million. In fact, the rate of growth in farm numbers during the last intercensus period—1997 to 2002—was greater for $5 million dollar farms (42 percent over the 5-year period) than for farms with sales from $1 million to $4,999,999 (35 percent), based on calculations from table 1.

Concentration of Production

The shift in farm production to million-dollar farms refl ects a long-term concentration of farm production on fewer farms that has been underway since at least the beginning of the 20th century (Hoppe, 2006). Farm policy debates about farm structure often focus on how soon the largest farms will dominate the production of commodities by commercial agriculture (Stanton, 1993).

Farming is not concentrated when compared with other U.S. industries, including those selling inputs to farmers and those buying farm products. There are still too many million-dollar farms (35,100) for an individual farm to hold much market power and dominate agriculture. Even when individual specializations are examined, production is not dominated by a few farms. Among million-dollar farms, for example, there are 7,100 high-value crop farms, 2,900 hog farms, 4,900 dairy farms, and 6,000 poultry farms.

Nevertheless, concentration may be approaching a level where it becomes a concern for specific commodities. The individual specializations used in ARMS for this analysis may include a variety of commodities, and ARMS data do not allow us to say much about concentration of individual commodities included in the specialization. The high-value crop specialization in particular includes a large number of specialty crops. Data on acres harvested from the 2002 Census of Agriculture suggest that production of some specialty crops occurs on a relatively small number of farms. For example, the 58 largest producers of head lettuce (out of 830 total producers) in 2002—each harvesting at least 1,000 acres of the crop—accounted for 65 percent of the total acreage in head lettuce. As another example, the 77 largest broccoli producers (out of 2,493 total producers)—each with at least 500 harvested acres of the crop—accounted for 69 percent of the total harvested acres.

Concern over the formation of monopolies or oligopolies does not become an issue in other industries until a small number of firms dominate the industry. For example, under the *Horizontal Merger Guidelines*—prepared by the U.S. Department of Justice and the Federal Trade Commission (1997) to evaluate proposed mergers—antitrust concerns would arise if a merger resulted in an industry of only four or five equal-sized firms (Kwoka and White, 2004). The same could be said for four firms accounting for 70 percent of the industry's production. By these standards, agricultural production is not concentrated, even if the number of farms dominating the production of a particular commodity falls to the point where it is measured in hundreds instead of thousands. Even lettuce and broccoli production would not currently be considered concentrated under the guidelines.

Concentration in agriculture can be more pronounced, however, if *ownership* of the commodity produced is considered, rather than farms producing it. Consider hogs as an example. Farmers will contract with a firm—often a large corporation such as a packing company—to take custody of the contractor's hogs and feed them out (Key and McBride, 2007). The contractor may own thousands of hogs located on multiple farms. Lawrence and Grimes (2007) estimate that 27 entities marketing (or removing) at least 500,000 hogs per year accounted for 43 percent of U.S. hog sales in 2006. But even in this case, the industry is not highly concentrated by the standards used in antitrust analyses.

The concentration of farm production has not elicited legislation to regulate the market power of farms or processors. However, the concentration of livestock production on fewer farms—dairy, hogs, poultry, and beef in the case of million-dollar farms—raises an environmental issue due to the manure associated with confined livestock production. Federal, State, and local governments have reacted with a variety of regulations (Ribaudo and Gollehon, 2006). The U.S. Environmental Protection Agency (EPA) introduced regulations in 2003 under the Clean Water Act to control the runoff of manure nutrients from the largest livestock feeding operations.

Million-Dollar Farms Are Family Farms

Million-dollar farms are overwhelmingly family operations. Eighty-four percent operate as family farms, and only 7 percent are organized as nonfamily corporations, generally with no more than 10 stockholders. The situation is similar for the largest million-dollar farms—those with sales of $5 million or more—although a smaller share are classified as family operations (64 percent) and a larger share as nonfamily corporations (21 percent), again generally with no more than 10 stockholders.

Direct ownership of million-dollar farms by large, publicly held nonfarm corporations is negligible, but somewhat more important for larger farms. Only 3 percent of the smaller million-dollar farms, those with sales less than $5 million, are part of a larger firm or corporation, while 10 percent of $5-million farms are part of a larger organization. Nevertheless, large nonfarm corporations are more involved in farming by acting as contractors. Some contractors—approximately 5,400 according to ARMS—are also farms. By 2006, 39 percent of production was under contract and million-dollar farms accounted for 62 percent of contract production. Further growth in contracting is still possible since some commodities—most grains, for example—are still largely sold in cash markets, while other commodities have not completely shifted to contracts (Hoppe et al., 2007). Much of the growth in contracting will occur on million-dollar farms or farms growing to that size. Large processors lower their transactions costs by establishing long-term relationships with large producers that secure a reliable flow of farm products at a volume allowing them to operate near full capacity and achieve economies of scale (MacDonald et al., 2000).

Although million-dollar farms are generally family operations, most of the labor (84 percent) is hired or contracted. The operator and spouse account for only about 10 percent of total labor hours worked. In contrast, the operator and spouse still account for 39 percent of the labor on farms with sales just under $1 million, in the $500,000 to $999,999 sales class. The heavy use of hired/contract labor on million-dollar farms simply reflects the size of the operations. The availability of farm labor, wage rates, and other labor issues is critical to production on these farms. Smaller farms—because they use more family labor—have greater independence from the local farm labor market.

REFERENCES

Ahearn, Mary; Banker David and Penni Korb. "*Farm Policies and the Evolving Organization of U.S. Family Farms*," paper for conference of the Société Française d'Economie Rurale, Paris, France, April 22-23, 2004.

Allen, Douglas W and Lueck Dean. "The Nature of the Farm," *Journal of Law and Economics,* Vol. 61, pp. 343-386, October 1998.

Allen, Rich. "*How to Interpret New Demographic Information in the Preliminary 2002 Census of Agriculture Release*," paper for 2004 Agricultural Outlook Forum, Arlington, VA, February 19-20, 2004. Available at: http://ageconsearch.umn.edu/bitstream/33028/1/fo04al02.pdf

Banker, David E and MacDonald, James M. (Eds.). *Structural and Financial Characteristics of U.S. Farms: 2004 Family Farm Report,* AIB-797, U.S. Department of Agriculture, Economic Research Service, March 2005.

Cochrane, Willard W. *The Development of American Agriculture: A Historical Analysis,* Minneapolis, MN: University of Minnesota Press, 1993.

DeNavas-Walt; Carmen; Proctor Bernadette D and Jessica Smith. *Income, Poverty, and Health Insurance Coverage in the United States: 2006,* P60-233, U.S. Department of Commerce, U.S. Census Bureau, August 2007.

Gray, Allan W and Michael Boehlje D. "*Drivers of Change in U.S. Agriculture: Implications for Future Policy*," paper for Joint AAEA/CAES Meetings, Portland, OR, July 2007.

Hoppe, Robert A. "Land Ownership and Farm Structure," *Agricultural Resources and Environmental Indicators, 2006 Edition,* Keith Wiebe and Noel Gollehon, (Eds.), EIB-16, U.S. Department of Agriculture, Economic Research Service, July 2006.

Hoppe, Robert A; Penni, Korb; O'Donoghue, Erik J and Banker, David E. *Structure and Finances of U.S. Farms: Family Farm Report, 2007 Edition*, EIB-24, U.S. Department of Agriculture, Economic Research Service, June 2007.

Hoppe, Robert A & Penni Korb. *Understanding U.S. Farm Exits*, ERR-21, U.S. Department of Agriculture, Economic Research Service, June 2006.

Hoppe, Robert A; Penni, Korb; Robert, Green; Mishra, Ashok and Sandretto, Carmen "Characteristics of Top-Performing Farms," *Structural and Financial Characteristics of U.S. Farms: 2004 Family Farm Report*, David E. Banker and James M. MacDonald, (Eds.), AIB-797, U.S. Department of Agriculture, Economic Research Service, March 2005.

Kwoka, John E Jr. and White, Lawrence J (Eds.). *The Antitrust Revolution: Economics, Competition, and Policy*, New York: Oxford University Press, 2004.

Lawrence, John D and Grimes, Glenn "*Production and Marketing Characteristics of U.S. Pork Producers, 2006*," ISU Economics Working Paper #070 14, Iowa State University, Department of Economics, Ames, Iowa, June 2007.

Key, Nigel and McBride, William *The Changing Economics of U.S. Hog Production*, ERR-52, U.S. Department of Agriculture, Economic Research Service, December 2007.

MacDonald, James M; O'Donoghue, Erik J; McBride, William D; Nehring, Richard F; Sandretto, Carmen L and Mosheim, Roberto *Profits, Costs, and the Changing Structure of Dairy Farming*, ERR-47, U.S. Department of Agriculture, Economic Research Service, September 2007.

MacDonald, James M; Korb, Penni and Hoppe, Robert "Experience Counts: Farm Business Survival in the U.S.," *Amber Waves*, Vol. 5, No.2, pp. 10-15, U.S. Department of Agriculture, Economic Research Service, April 2007.

MacDonald, James M and Banker, David E. "Agricultural Use of Production and Marketing Contracts," *Structural and Financial Characteristics of U.S. Farms: 2004 Family Farm Report*, David E. Banker and James M. MacDonald, (Eds.), AIB-797, U.S. Department of Agriculture, Economic Research Service, March 2005.

MacDonald, James M; Ollinger, Michael E; Nelson, Kenneth E and Handy, Charles R *Consolidation in U.S. Meatpacking*, AER-785, U.S. Department of Agriculture, Economic Research Service, February 2000.

McBride, William D and Nigel Key. *Economic and Structural Relationships in U.S. Hog Production,* AER-8 18, U.S. Department of Agriculture, Economic Research Service, February 2003.

Ribaudo, Marc and Gollehon, Noel "Animal Agriculture and the Environment," *Agricultural Resources and Environmental Indicators, 2006 Edition,* Keith Wiebe and Noel Gollehon, (Eds.), EIB-16, U.S. Department of Agriculture, Economic Research Service, July 2006.

Stanton, BF "Farm Structure: Concept and Definition," *Size, Structure, and the Changing Face of American Agriculture,* Arne Hallam, (Ed.), Boulder, CO: Westview Press, 1993.

U.S. Department of Agriculture, National Agricultural Statistics Service. Agricultural Statistics 2007, July 2007.

———. *2002 Census of Agriculture,* Vol. 1: Geographic Area Series, Part 51: United States Summary and State Data, AC-02-A-5 1, June 2004.

U.S. Department of Agriculture, National Commission on Small Farms. *A Time to Act: A Report of the USDA National Commission on Small Farms,* Miscellaneous Publication 1545 (MP-1545), January 1998.

U.S. Department of Commerce, Bureau of the Census. *1982 Census of Agriculture,* Vol. 2, Subject Series, Part 2: Coverage Evaluation, AC82-SS-2, April 1985.

U.S. Department of Justice and the Federal Trade Commission. *Horizontal Merger Guidelines,* revised April 8, 1997

U.S. Department of the Treasury, Internal Revenue Service. *Tax Issues for Limited Liability Companies,* Publication 3402, revised March 2008

APPENDIX: THE 2002 CENSUS OF AGRICULTURE LONGITUDE FILE

Data from the 2002 Census of Agriculture Longitudinal File are used in this report to trace the history of million-dollar farms that existed as of 2002. The 2002 file was created by updating the 1997 Census of Agriculture Longitudinal File—which contained data from five previous censuses (1978, 1982, 1987, 1992, and 1997)—with data from the 2002 Census of Agriculture. As a result, individual farm businesses can be tracked from 1978 to 2002. This appendix presents a brief overview of the 2002 Census of Agriculture Longitudinal File. For more detailed information about how longitudinal files are built from the census of agriculture, see Hoppe and Korb (2006) and MacDonald et al. (2007).

Linking Censuses of Agriculture

The 2002 longitudinal file links records from each census for individual farms, using an identification number (ID). The ID identifies a farm operation for a particular census and follows the farm operation through subsequent censuses (up to six). Because continuing farm businesses retain the same ID—while new farm businesses receive new ones—a farm's record for each census can be linked. A farm is defined as going out of business when there is no response to the census questionnaire or the questionnaire is returned with a statement that the establishment is no longer operating as a farm. A farm that has gone out of business (or exited) is coded with a zero in the ID variable field for the year of exit. A farm operation whose ID cannot be matched or linked to a previous record would be considered a new business (an entry) and added to the longitudinal file as a new record.

The longitudinal file follows farm businesses, rather than farm operators. Thus, an operation changing hands does not necessarily mean that the original farm went out of business and a new farm appeared on the longitudinal file. For example, a widow or adult child assuming the operation of the farm upon the death of the operator would not count as an exit. Selling the farm to an unrelated operator, who continues the business as a separate entity, is also not an exit. The operator and farm may not exit together. A common example of dual exit, however, occurs when the farm operator stops farming and rents or sells the land to other farmers who incorporate it into existing operations.

Appendix Table 1. Business Age Classes

Business age class	Census when farm first appears	Year of entry
Less than 5 years	2002	Between 1998 and 2002
5 to 9 years	1997	Between 1993 and 1997
10 to 14 years	1992	Between 1988 and 1992
15 to 19 years	1987	Between 1983 and 1987
20 to 23 years	1982	Between 1979 and 1982
24 years or more	1978	1978 or earlier

Source: Hoppe and Korb (2006) and MacDonald et al. (2007).

Business Age

Business age on the longitudinal file is based on which census the farm first appears. For example, a farm that appears in the 2002 Census of Agriculture may have entered farming as early as 1998 (immediately after the 1997 Census), but no later than 2002. The farms age would be reported as less than 5 years old as of 2002. Using similar logic, five additional age classes were created, one for each of the remaining census used in the longitudinal file. The six age classes used in this report are outlined in appendix table 1.

Limitations of the Data

The longitudinal file is not truly longitudinal, like the University of Michigan's Panel Study of Income Dynamics (PSID), which was designed to follow households over time. Rather than identifying farms and following them as time progresses, the longitudinal file links data collected in the past for another purpose (the agricultural census). Because the census of agriculture is not designed to track businesses over time, errors linking records in the longitudinal file may lead to an overstatement of exits and entrances. Nevertheless, analysis of the 1997 longitudinal file, predecessor to the 2002 file, shows U.S. farm exits are similar to those in other industries and countries (Hoppe and Korb, 2006).

One problem linking observations across multiple censuses is "whole farm nonresponse," when an operator does not respond to a census after numerous attempts. Some farms classified as exits may have been continuing operations that failed to respond to the census questionnaire. Similarly, some farms classified as entries may be continuing operations that did not respond to the previous census. Nonresponse, however, may be less of an issue for million-dollar farms due to intensive efforts to get responses from large or unique operations (USDA, NASS, 2004).

End Notes

[1] Differences between ARMS-based estimates are generally stressed in this report only when the estimates are significantly different at the 95-percent confidence level or more.

[2] Sales are defined here to include the gross market value of agricultural products sold or removed from farms, before taxes and production expenses (USDA, NASS, 2002).

Government payments are excluded from sales in figure 1, since data on these payments were not collected prior to the 1987 census.

[3] The U.S. Department of Agriculture defines a farm as any place that produced and sold or normally would have produced and sold $1,000 worth of agricultural products during the year. If a place did not have $1,000 in sales, a "point system" assigns values for acres of various crops and head of livestock to estimate normal sales. "Point farms" are farms with less than $1,000 in sales but earn points worth at least $1,000. See "What is the Definition of a Farm?" on the NASS website at: http://www. agcensus.usda.gov/Help/FAQs/2002_Census/index.asp#1.

[4] Constant-dollar sales classes cannot be prepared before 1982 due to incomplete census records for individual farms. A computer file with individual farm observations is available for the 1978 Census of Agriculture, but these observations cannot be weighted to U.S. totals (Hoppe and Korb, 2006) and were excluded from figure 1 and table 1. Data for 1978 were included in table 2, however, because it was not necessary to weight 1978 observations to U.S. totals to determine the business age of farms existing in 2002

[5] Gross farm sales in ARMS include Government payments received by the farm and its share landlords. We did not remove these payments from ARMS sales—to be consistent with the measure of sales used in the census data—because we wanted to examine the receipt of Government payments among million- dollar farms. Removing Government payments from sales would have reduced the ARMS count of million-dollar farms by only 5 percent in 2006.

[6] The value of production measures crops and livestock produced in a given year and excludes the effects of inventory changes (unlike gross farm sales). It is calculated by multiplying the quantity of commodities grown by the price of the commodity. In some cases, quantities are not available from ARMS, and cash sales are used as a proxy for price multiplied by the quantity. These cases generally involve perishable commodities, such as high-value crops, livestock, and livestock products sold without a contract. Sales from inventory are less of an issue for perishable commodities. Note that the value of production excludes the value of crops grown to feed livestock on the same farm.

[7] The difference between the 3-percent estimate for smaller million-dollar farms and the 10-percent estimate for $5-million farms is statistically signifi- cant only at the 90 percent level.

[8] Marginally solvent farms have positive net farm income and a debt/asset ratio greater than 40 percent, while vulnerable farms have negative net farm income and a debt/asset ratio greater than 40 percent. See footnote six of table 10 for more information.

[9] The income estimates discussed in this section are for the household of the principal operator of a farm. Any income received by the households of secondary operators is excluded.

[10] Different versions of the ARMS are conducted each year to collect information useful for specific purposes. All five versions of the 2006 ARMS collected the number of hours worked on farms by the principal operator, the spouse of the principal operator, other operators, and unpaid workers. (ARMS does not differentiate between operators' management and labor hours.) Version 1 of the survey also collected the number of the hours worked by hired laborers. Hours of hired labor on the other versions were estimated by dividing cash wages for hired labor by the State-specific wage rate for farm labor. No versions of the survey collected hours of contract labor, so an estimate was made by dividing contract labor expense by the State wage rate.

[11] For farms with production contracts, only the fees—rather than the value of the commodities removed—are included in gross cash farm income. The value of commodities removed, however, is included in sales. Measuring farm size by gross cash income rather than sales would reduce the number of million- dollar farms among some specializations, such as poultry farms (Hoppe et al., 2007).

[12] Discussions of economies of scale in dairy and hog production appear in MacDonald et al. (2007) and Key and McBride (2007).

In: Million Dollar Farms in the New Century ISBN: 978-1-60741-755-2
Editors: Samuel D. Bosworth © 2010 Nova Science Publishers, Inc.

Chapter 2

UNITED STATES FARM INCOME

Randy Schnepf

SUMMARY

Despite high production costs, 2008 represented another year of record profitability for the U.S. farm economy as a whole. According to USDA's Economic Research Service (ERS), national net farm income—a key indicator of U.S. farm well-being—rose to a record $86.9 billion in 2008, marginally above the previous year's record ($86.8 billion). The growth in cash receipts to a record $323.4 billion for crop and livestock sales (up $38.6 billion or 14% from 2007) was nearly offset by a surge in production costs (up $38.2 billion or 15%) to a record $292.5 billion.

Government farm payments are projected up slightly in 2008 at $12.5 billion. An increase in ad hoc disaster payments more than offset lower commodity program payments, which declined when high crop prices rose above the price triggers for marketing loan benefits and counter- cyclical payments in 2008.

Within the farm balance sheet, total farm asset value of $2,359 billion and total farm debt of $212 billion are both projected at record levels in 2008. The debt-to-asset ratio of 9.0% is down sharply from last year's value of 9.6% and represents the lowest level since 1960, suggesting a strong financial position for the agricultural sector as a whole.

However, less than ideal market conditions heading into 2009 suggest dim prospects for the longer-term farm income outlook, albeit surrounded by considerable uncertainty. On the one hand, the global financial crisis, economic recession, rising unemployment, limited credit availability, and plummeting asset values that persist in early 2009 have contributed to substantial "demand destruction" (i.e., a severe weakening of consumer demand), which bodes poorly for farm commodity price prospects. On the other hand, weak energy markets and declining input prices could provide some spark to both producer investment and consumer demand for agricultural sector products, perhaps by the middle to latter half of the year. USDA will release its first U.S. farm income forecasts for 2009 on February 12, 2009.

This report supersedes CRS Report RS2 1970, *The U.S. Farm Economy*, by Randy Schnepf. It will be updated as events warrant.

INTRODUCTION

The U.S. farm sector is vast and varied. It encompasses production activities related to traditional field crops (such as corn, soybeans, wheat, and cotton), livestock and poultry products (including meat, dairy, and eggs), as well as fruits, tree nuts, and vegetables. In addition, U.S. agricultural output includes greenhouse and nursery products, forest products, custom work, machine hire, and other farm-related activities. The intensity and economic importance of each of these activities, as well as their underlying market structure and production processes, vary regionally based on the agro-climatic setting, market conditions, and other factors. As a result, farm income and rural economic conditions may vary substantially across the United States.[1] However, this report focuses singularly on aggregate national net farm income and the farm debt-to-asset status as reported by the U.S. Department of Agriculture (USDA).[2]

Annual U.S. net farm income is the single most watched indicator of farm sector well-being, as it captures and reflects the entirety of economic activity across the range of production processes, input expenses, and marketing conditions that have persisted during a specific time period. When national net farm income is reported together with a measure of the national farm debt-to-asset situation, the two summary statistics provide a quick indicator of the economic well-being of the national farm economy.

Two different indicators measure farm profitability: net cash income and net farm income.

- **Net cash income** compares cash receipts to cash expenses. As such, it is a cash flow measure representing the funds that are available to farm operators to meet family living expenses and make debt payments. For example, crops that are produced and harvested but kept in on-farm storage are not counted in net cash income.

- **Net farm income** is a value of production measure, indicating the farm operator's share of the net value added to the national economy within a calendar year, independent of whether it is received in cash or noncash form. In contrast to net cash income, net farm income includes the value of home consumption, changes in inventories, capital replacement, and implicit rent and expenses related to the farm operator's dwelling that are not reflected in cash transactions during the current year.

Net cash income is generally less variable than net farm income. Farmers can manage the timing of crop and livestock sales and of the purchase of inputs to stabilize the variability in their net cash income. For example, farmers can hold crops from large harvests to sell in the forthcoming year, when output may be lower and prices higher. Off-farm income, which has increased in importance in recent decades, is not included in the calculation of aggregate farm income. Instead, it is included in the discussion of farm income at the household level.

REVIEW OF CALENDAR YEAR 2008

Despite the dramatic price declines that occurred during the second half of 2008 (**Figure 2** and **Figure 3**), USDA's net farm income estimate for 2008 remains at the historically high level of $86.9 billion, up marginally from the previous year's record $86.8 billion (**Table 1**).[3] When measured in cash terms, net cash income in 2008 is projected up nearly 4% to $90.7 billion, compared with the previous year's record of $87.4 billion (**Figure 1**). Net cash income is projected to rise more than net farm income because of the carryover of unsold crops from 2007 for sale in 2008.

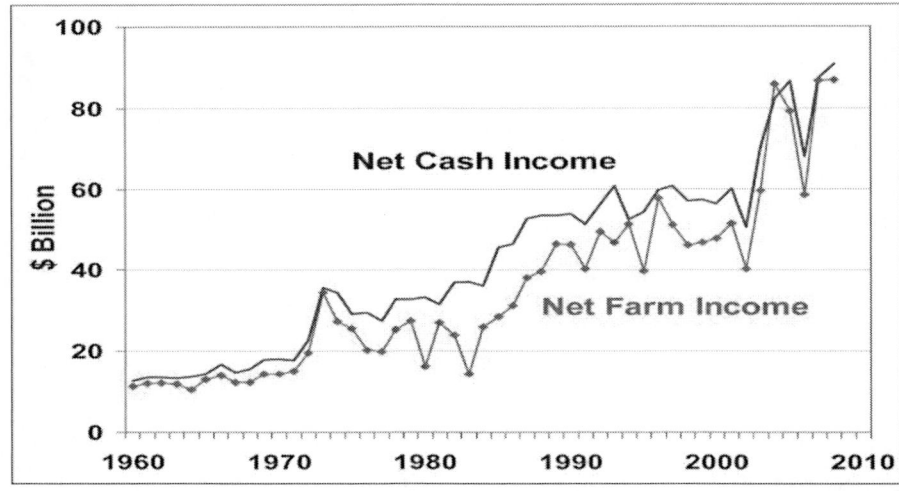

Source: USDA, Economic Research Service, "2008 Farm Income Forecast," November 25, 2008, at *http://www.ers.usda.gov/* Briefing/FarmIncome/.

Notes: All values are in nominal terms, i.e., they are not adjusted for inflation. 2008 is projected.

Figure 1. Annual U.S. Farm Sector Income, 1960 to 2008F

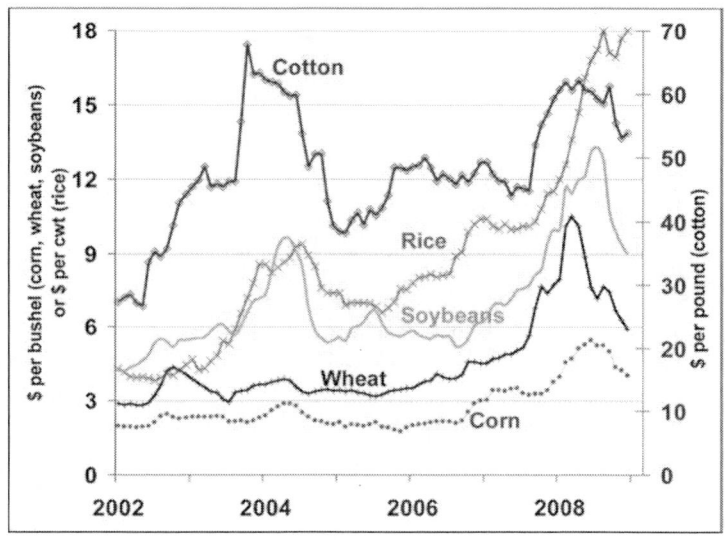

Source: USDA, National Agricultural Statistics Service.

Figure 2. Monthly Farm-Prices-Received for Major Field Crops

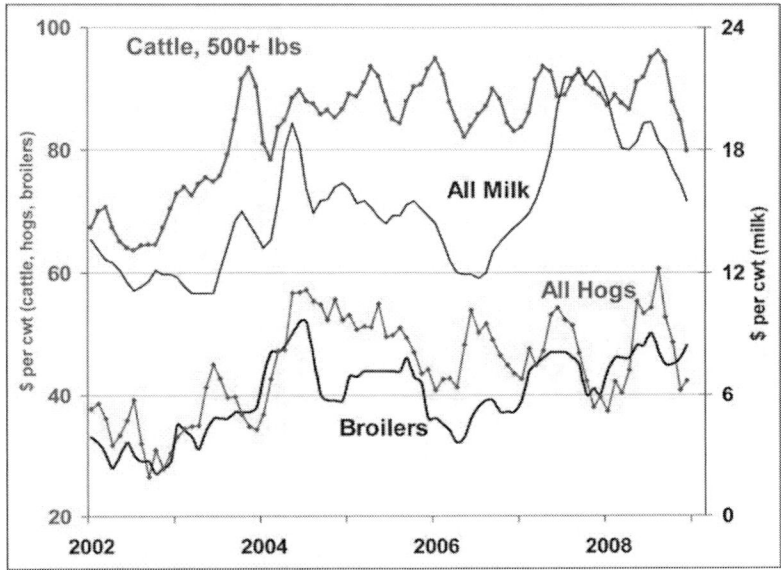

Source: USDA, National Agricultural Statistics Service.
Note: cwt = hundredweight or units of 100 lbs.

Figure 3. Monthly Farm-Prices-Received for Major Livestock Products

From a historical perspective, the six years extending from 2003 through 2008 represent the six highest years for U.S. farm income on record. National net cash income averaged nearly $81 billion per year during that six-year period, well above the previous single year high of $60.8 billion achieved first in 1993 and again in 1997. Prospects for 2009 are less sanguine, as commodity prices continue to weaken for most major field crops and livestock products heading into 2009 (**Figure 2** and **Figure 3**). Only rice and broilers appear to have reversed, at least temporarily, the downward trend.

Cash Receipts

Good harvests, strong prices, and robust domestic and international demand combined to push farm gross receipts to record levels in 2008. The combined value of cash receipts from sales of both crop and livestock commodities is projected at $323.4 billion in 2008, the highest amount on record—up $38.6 billion from the previous year's record of $284.9 billion—and driven almost entirely by higher crop prices (**Table 1** and **Figure 4**).

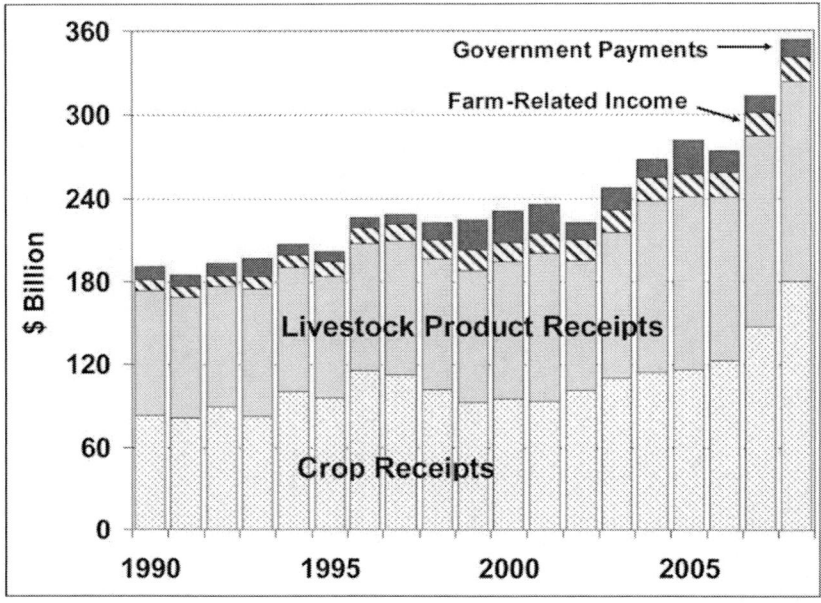

Source: USDA, Economic Research Service, "2008 Farm Income Forecast," November 25, 2008, at

Notes: 2008 is projected. Receipts from crop and livestock product sales, as well as government payments, are described in more detail below. Farm-related income includes income from custom work, machine hire, agritourism, forest product sales, insurance indemnities, and cooperative patronage dividend fees.

Figure 4. Farm Cash Receipts by Source

Table 1. Annual U.S. Farm Income Since 2002 ($ billions)

Item	2002	2003	2004	2005	2006	2007	2008[a]
1. Cash receipts	195.0	215.6	237.2	240.9	240.8	284.8	323.4
Crops [b]	101.0	109.9	113.6	116.0	122.6	147.0	179.9
Livestock [c]	94.0	105.6	123.6	124.9	118.2	137.9	143.5
2. Government payments [d]	12.4	16.5	13.0	24.4	15.8	11.9	12..5
Fixed direct payments [e]	3.9	6.4	5.2	5.2	5.1	5.1	5.2
CCP [f]	0.2	2.3	1.1	4.1	4.0	1.1	0.7
Marketing Loan Benefits	2.8	1.3	3.5	7.1	1.8	1.1	0.1
Conservation	2.0	2.2	2.3	2.8	3.0	3.1	3.2
Ad hoc and emergency	1.7	3.1	0.6	3.2	0.3	0.5	2.7
All otherg [h]	1.9	1.2	0.2	2.1	1.7	1.0	0.6

Item	2002	2003	2004	2005	2006	2007	2008[a]
3. Farm-related income	14.8	15.7	17.1	16.2	17.5	16.6	17.6
4. Gross cash income (1+2+3)[i]	222.2	247.8	267.3	281.5	274.1	313.4	353.5
5. Cash expenses	171.6	177.6	185.0	194.8	206.0	226.0	262.8
6. NET CASH INCOME	50.7	70.2	82.3	86.6	68.0	87.4	90.7
7. Total gross revenues[j]	233.6	260.0	295.6	301.1	292.4	341.1	379.4
8. Total production expenses [k]	193.4	200.3	209.8	221.8	233.9	254.4	292.5
9. NET FARM INCOME	40.2	59.7	85.8	79.3	58.5	86.8	86.9

Source: USDA, Economic Research Service, briefing rooms: Farm Income and Costs: Farm Sector Income, and Costs: Farm Sector Income. U.S. farm income data updated as of November 25, 2008.

a. Data for 2008 are USDA forecasts.
b. Includes Commodity Credit Corporation loans under the farm commodity support program.
c. Government payments reflect payments made directly to all recipients in the farm sector, including landlords. The non-operator landlords' share is offset by its inclusion in rental expenses paid to these landlords and thus is not reflected in net farm income or net cash income. For more information on U.S. farm commodity programs, see CRS Report RL34594, Farm Commodity Programs in the 2008 Farm Bill, by Jim Monke; for more information on conservation programs see CRS Report RL34557, Conservation Provisions of the 2008 Farm Bill, by Tadlock Cowan, Renée Johnson, and Megan Stubbs.
d. Direct payments include production flexibility payments of the 1996 Farm Act through 2001, and fixed direct payments under the 2002 Farm Act since 2002.
e. CCP = counter-cyclical payments.
f. Includes loan deficiency payments (LDP); marketing loan gains (MLG); and commodity certificate exchange gains.
g. Peanut quota buyout, milk income loss payments, and other miscellaneous program payments.
h. Income from custom work, machine hire, agri-tourism, forest product sales, and other farm sources.
i. Excludes depreciation and perquisites to hired labor.
j. Gross cash income plus inventory adjustments, the value of home consumption, and the imputed rental value of operator dwellings .
k. Cash expenses plus depreciation and perquisites to hired labor.

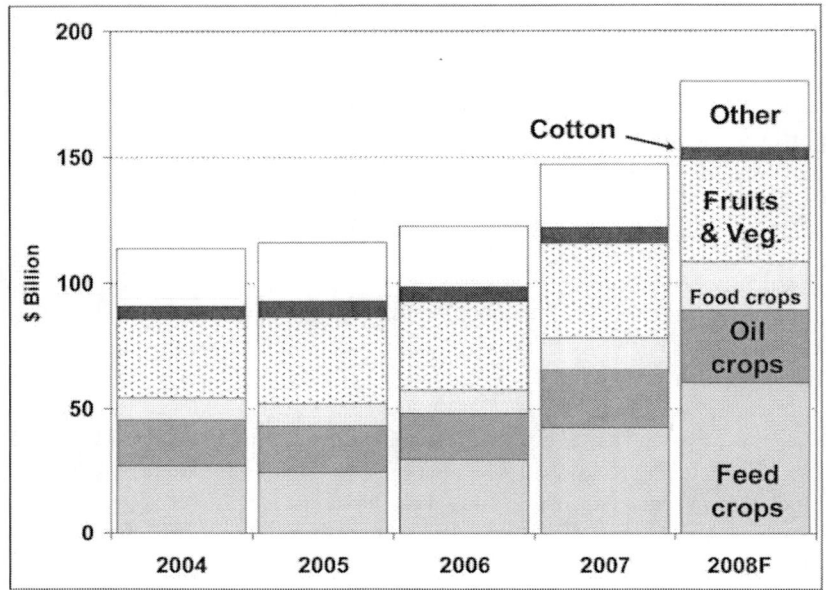

Source: USDA, Economic Research Service, "2008 Farm Income Forecast," November 25, 2008, at http://www.ers.usda.gov/Briefing/FarmIncome/.
Notes: 2008 is projected.

Figure 5. Crop Cash Receipts by Source

In recent years, U.S. domestic demand has been supported by the rapid emergence of the U.S. agriculture-based ethanol industry. In addition, strong export demand through 2007 and the first half of 2008, aided in part by a weak dollar, helped to draw stocks for major grains and oilseeds to historically low levels in 2008, thus supporting higher market prices. While crop farmers rejoiced, livestock feeders expressed concern about the escalating costs of feed—the largest single cost component for cattle, dairy, hog, and poultry production.

Crops

Strong commodity prices, although highly volatile since mid-2008, are expected to generate record crop cash receipts in 2008 of $180 billion (**Figure 5**). Sales of field crops (i.e., feed, food, and oil crops) increased over 29% from 2007 to $113.3 billion. This total includes feed crop (i.e., corn, sorghum, barley, and oats) sales of $60.2 billion, up 42%, food crop (i.e., wheat and rice) sales of $19.2 billion, up over 50%, and oil crop (i.e., soybeans, sunflowers, rapeseed/canola, and other minor oilseeds) sales of $28.7 billion, up 27%. Corn cash receipts alone are projected at a record $50 billion in 2008—the single most valuable farm crop ever harvested in the United States. Cash receipts for soybeans ($26.6 billion), wheat ($16.1 billion), and rice ($3.1 billion) also are expected to establish single-season sales value records. Cotton was the most notable deviation from this trend, with a sales decline of 17% to $5.1 billion, due to sharply lower plantings and output in 2008.

Table 2. Average Annual Income per U.S. Household, Farm versus All, 2002-2008

	2002	2003	2004	2005	2006	2007	2008
	($ per household)						
Average U.S. Farm Income by Source							
On-Farm Income	$3,477	$7,884	$13,564	$13,996	$8,750	$8,605	$5,901
Off-Farm income	$62,284	$60,713	$67,279	$67,091	$72,502	$77,618	$80,897
Total Farm income	$65,761	$68,597	$80,843	$81,086	$81,251	$86,223	$86,798
Average U.S. Household Income	$57,852	$59,067	$60,466	$63,344	$66,570	$67,609	na
	(percent)						
Farm Household Income as Share of U.S. Average Household Income	114%	116%	134%	128%	122%	128%	na

Source: USDA, ERS Briefing Room: Farm Household Economics and Well-Being: Historic Data On Farm Operator Household Income, at http://www.ers.usda.gov/Briefing/WellBeing/Gallery/historic.htm.

Table 3. Average Annual Farm Sector Debt-to-Asset Ratio, 2002-2008F

	2002	2003	2004	2005	2006	2007	2008
	($ billions)						
Farm Assets	$1,304.0	1,378.8	1,617.6	1,835.5	2,047.4	2,209.9	2,35
Farm Debt	193.3	175.1	183.0	193.2	196.4	211.5	211.7
Farm Equity	1,110.7	1,203.6	1,434.6	1,642.2	1,851.0	1,998.4	2,14
	(percent)						
Debt-to-Asset Ratio	14.8%	12.7%	11.3%	10.5%	9.6%	9.6%	9.0%

Source: USDA, ERS Briefing Room: Farm Household Economics and Well-Being: Farm Business Balance Sheet

Note: 2008 is projected.

Table 4. U.S. Prices and Support Rates for Selected Farm Commodities Since 2002

Commodity a	Unit	Year	2002/03	2003/04	2004/05	2005/06	2006/07	2007/08	2008/09Fb	% Change	2009/10F	Loan rated	Target Price
Wheat	$/bu	Jun-May	3.56	3.40	3.40	3.42	4.26	6.48	6.50-6.90	3.4%	—	2.75	3.92
Corn	$/bu	Sep-Aug	2.32	2.42	2.06	2.00	3.04	4.20	3.565-4.25	-4.8%	—	1.95	2.63
Sorghum	$/bu	Sep-Aug	2.32	2.39	1.79	1.86	3.29	4.08	2.90-3.50	-19.1%	—	1.95	2.57
Barley	$/bu	Jun-May	2.72	2.83	2.48	2.53	2.85	4.02	4.95-5.35	46.8%	—	1.85	2.44
Oats	$/bu	Jun-May	1.81	1.48	1.48	1.63	1.87	2.63	2.90-3.10	10.3%	—	1.33	1.44
Rice	$/cwt	Aug-Jul	4.49	8.08	7.33	7.65	9.96	12.80	16.50-17.50	22.3%	—	6.50	10.50
Soybeans	$/bu	Sep-Aug	5.53	7.34	5.74	5.66	6.43	10.10	8.50-9.50	-10.9%	—	5.00	5.80
Soybean oil	¢/lb	Oct-Sep	22.0	30.0	23.0	23.4	31.0	52.0	32.0-35.0	-36.6%	—	—	—
Soybean meal	$/st	Oct-Sep	181.6	256.1	182.9	174.2	205.4	335.9	250-310	-19.6%	—	—	—
Cotton, Upland	¢/lb	Aug-Jul	44.5	61.8	41.6	47.7	46.5	59.3	44.0-52.0	-22.4%	—	52.00	71.25
Choice Steers	$/cwt	Jan-Dec	67.0	84.7	84.8	87.3	85.4	91.8	92.27	0.8%	91-97	—	—
Barrows/Gilts	$/cwt	Jan-Dec	34.9	39.5	52.5	50.1	47.3	47.1	47.84	1.3%	47-51	—	—
Broilers	¢/lb	Jan-Dec	55.6	62.0	74.1	70.8	64.4	76.4	79.7	4.3%	81-87	—	—
Eggs	¢/doz	Jan-Dec	67.1	87.9	82.2	65.5	71.8	114.4	127.7	11.6%	118-127	—	—
Milk		Jan-Dec	12.11	12.52	16.05	15.14	12.90	19.13	18.30-18.40	-4.1%	11.80-	—	—

Source: Various USDA agency sources as described in the notes below.

a. Season average farm price for grains and oilseeds are from USDA, National Agricultural Statistical Service, Agricultural Prices. Calendar year data is for the first year, e.g., 2000/2001 = 2000; F = forecast from World Agricultural Supply and Demand Estimates (WASDE) January 12, 2009; — = no value ; and USDA's out-year 2009/2010 crop price forecasts will first appear in the May 2009

WASDE report. Soybean and livestock product prices are from USDA, Agricultural Marketing Service (AMS): soybean oil—Decatur, IL, cash price, simple average crude; soybean meal—Decatur, IL, cash price, simple average 48% protein; choice steers—Nebraska, direct 1100-1300 lbs.; barrows/gilts—national base, live equivalent 51%-52% lean; broilers—wholesale, 12-city average; eggs—Grade A, New York, volume buyers; and milk—simple average of prices received by farmers for all milk.

b. Data for 2008/09 and 2009/10 are USDA forecasts.

c. Percent change from 2007/08, calculated using the difference from the midpoint of the range for 2008/09 with the estimate for 2007/08.

d. Loan rate and target prices are for the 2008/09 crop year. For more information, see CRS Report RL34594, *Farm Commodity Programs in the 2008 Farm Bill*, by Jim Monke.

Crop-specific output and sales are increasingly being influenced by the rapid expansion of corn- based biofuels production, due in large part to strong federal incentives that include a blending tax credit, a usage mandate, and a prohibitive import tariff on foreign-produced biofuels. With this strong support, the U.S. corn-based ethanol industry has grown rapidly since mid-2004, when production capacity was estimated at around 3 billion gallons per year, to over 10.8 billion gallons as of December 30, 2008.[4] The U.S. ethanol sector received a substantial boost in December 2007 when the Energy Independence and Security Act (EISA) was signed into law (P.L. 110-140). EISA greatly expands the mandate for corn-based ethanol use from 4.7 billion gallons in 2007 to 9 billion in 2008 and 15 billion by 2015.[5] USDA estimates that nearly 30% of the 2008 corn crop will be used to produce ethanol during the 2008/2009 (September-August) corn marketing year.[6] This additional demand has helped to push corn and other crop prices steadily higher since 2005 as they compete for a fixed amount of cropland (Table 4).

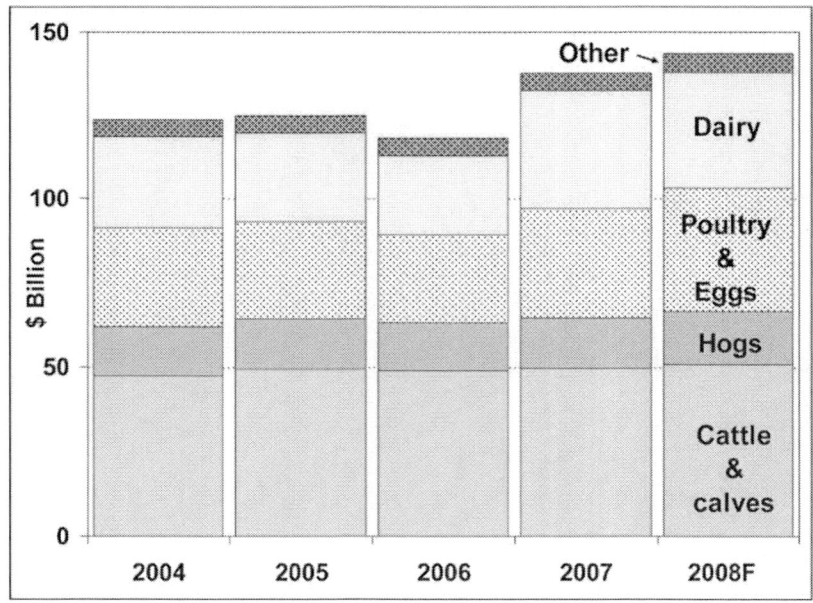

Source: USDA, Economic Research Service, "2008 Farm Income Forecast," November 25, 2008, at http://www.ers.usda.gov/Briefing/FarmIncome/.
Notes: 2008 is projected.

Figure 6. U.S. Livestock Product Cash Receipts by Source

Livestock

As with crop sales, the value of livestock product sales also is forecast record high in 2008 at $143.5 billion, up 4% from the previous year's record of $137.9 billion (Figure 6).

Government Payments Strong market prices for most major livestock categories—beef, poultry, and hogs—during 2008 were the driving factor behind record cash receipts for livestock products. However, price prospects have weakened considerably heading into 2009—milk prices are projected about 34% lower in 2009 and eggs down about 4% (**Table 4**).

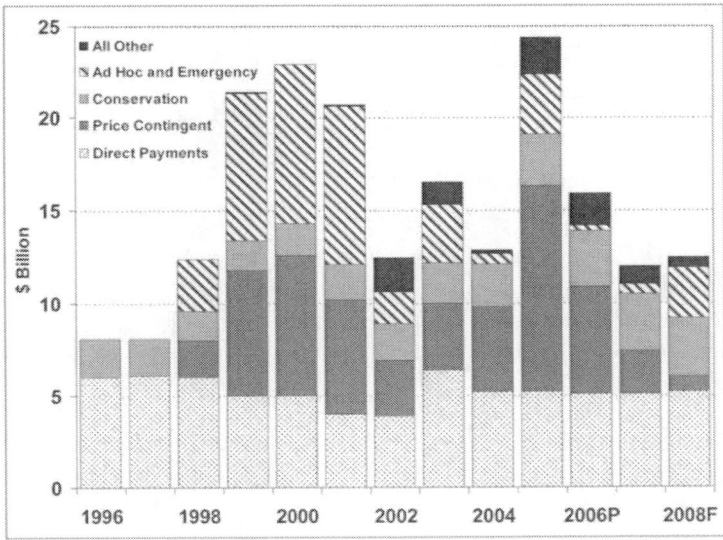

Source: USDA, Economic Research Service, "2008 Farm Income Forecast," November 25, 2008, at http://www.ers.usda.gov/Briefing/FarmIncome/; and USDA, Risk Management Agency, Current Year-to-Date National Summary of Business Reports, at *http://www.rma.usda.gov/data/sob.html*.

Notes: Data are on a fiscal year basis and may not correspond exactly with the crop or calendar year; 2008 is projected. Direct payments include production flexibility contract payments enacted under the 1996 farm bill and fixed direct payments of the 2002 and 2008 farm bills; price-contingent outlays include loan deficiency payments (LDPs), marketing loan gains, and counter-cyclical payments (CCPs); conservation outlays include Conservation Reserve Program (CRP) payments along with other conservation program outlays; Ad Hoc and Emergency includes emergency supplemental crop and livestock disaster payments and market loss assistance payments for relief of low commodity prices; and "all other" outlays include peanut quota buyout payments, milk income loss payments, tobacco transition payments, and other miscellaneous expenditures.

Figure 7. U.S. Government Farm Support, Direct Outlays, 1996 to 2008F

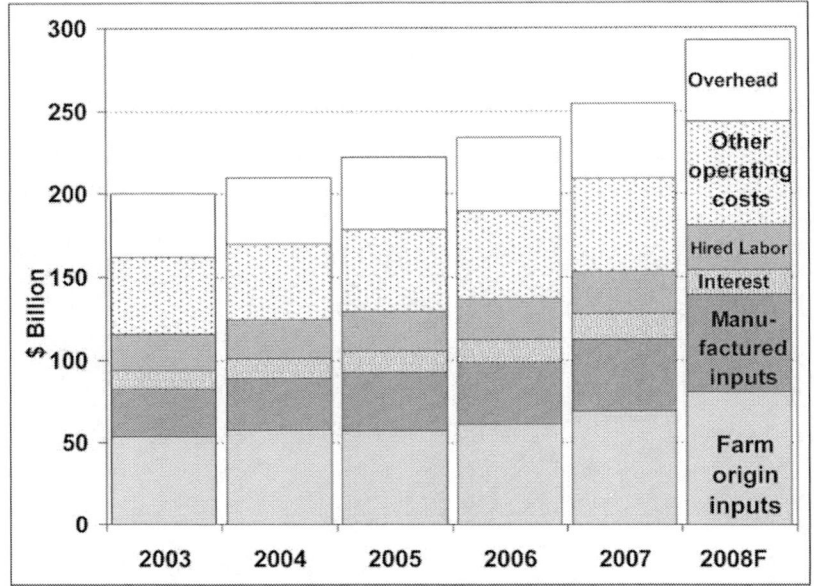

Source: USDA, Economic Research Service, "2008 Farm Income Forecast," November 25, 2008, at http://www.ers.usda.gov/Briefing/FarmIncome/.

Notes: 2008 is projected. Farm origin inputs include purchases of feed, livestock and poultry, and seed. Manufactured inputs include fertilizers and lime, pesticides, petroleum fuel and oils, and electricity. Other operating costs include repair and maintenance of capital items, machine hire and custom work, marketing storage, transportation expenses, and other miscellaneous expenses. Overhead expenses include property taxes, net rent to a non-operator landlord, and capital consumption.

Figure 8. Farm Cash Production Expenses by Source

Government Payments

Direct government payments are forecast at $12.5 billion in 2008, up about 5% from $11.9 billion in 2007 but well below the record of $24.4 billion in 2005. Higher market prices during 2008 limited payments under the two major price-contingent programs—counter-cyclical payments (CCP), which are projected to decline to about $0.7 billion, and marketing loan benefits (loan deficiency payments, marketing loan gains, and certificate exchange gains), which are projected to fall to about $90 million in 2008.[7] In contrast, farm disaster assistance—which (along with other emergency assistance) has figured heavily in farm sector income in most of the previous 20 years (1989-2008)[8]—is projected up at $3.2 billion in 2008.

Fixed direct payments, whose payment rates are fixed in legislation and are not affected by the level of program crop prices, have been $5.1 to $5.2 billion since 2004 and will only vary based on payment timing and participation rates. Conservation payments have grown slowly but steadily since 1998, and are expected to reach $3.2 billion in 2008.

Production Expenses

An emerging concern for the U.S. farm sector has been the rapid escalation of total farm production expenses (Figure 8 and Figure 9), which have risen an average of 7.8% annually since 2003 compared with an average annual growth rate of only 7.1% for gross cash income.

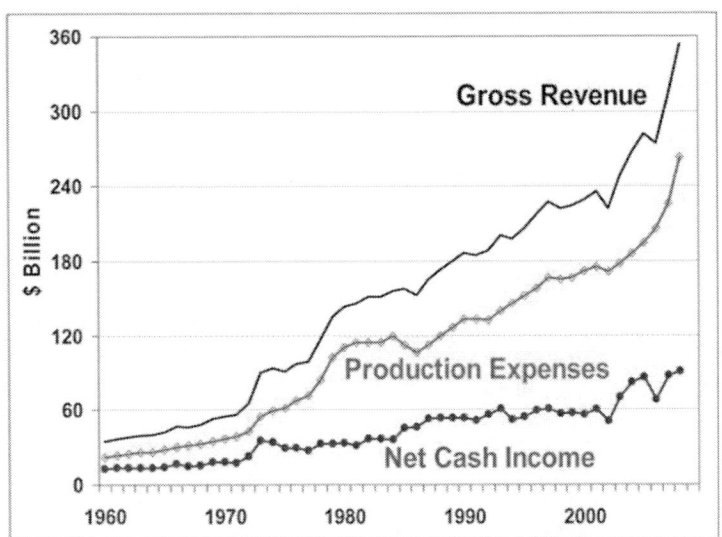

Source: USDA, Economic Research Service, "2008 Farm Income Forecast," November 25, 2008, at http://www.ers.usda.gov/Briefing/FarmIncome/.
Notes: All values are in nominal terms, i.e., not adjusted for inflation. 2008 is projected.

Figure 9. U.S. Farm Gross Revenue, Production Expenses, and Net Income (cash basis)

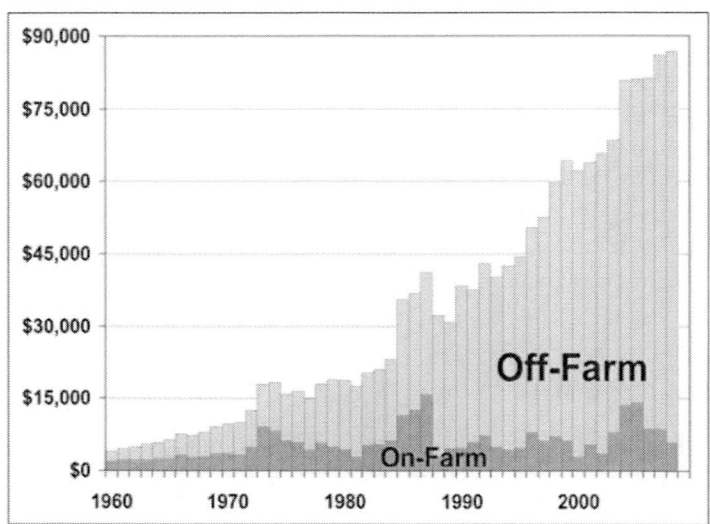

Source: USDA, Economic Research Service, "2008 Farm Income Forecast," November 25, 2008, at [http://www.ers.usda.gov/Briefing/FarmIncome/].
Notes: 2008 is projected.

Figure 10. U.S. Average Farm Household Income, by On- and Off-Farm Sources

Total production expenses are forecast at a nominal record $262.8 billion in 2008, up $36.8 billion (16%) from 2007's record level (**Figure 8**). Higher prices for feed crops (although beneficial for feed crop producers) have raised feed costs nearly 23% year-to-year for livestock producers. Feed costs are expected to rise to $47 billion in 2008, while escalating energy costs push fuels and fertilizers up 26% and 64% year-over-year, respectively, to record outlays of $16.4 and $27.5 billion in 2008. However, both feed and energy costs declined sharply in the last quarter of 2008, and appear likely to remain at substantially lower levels in 2009.

AVERAGE FARM HOUSEHOLD INCOME

Average farm household income is projected at a record $86,798 in 2008, up 0.7% from the previous year's record. The share of farm income derived from off-farm sources has been increasing steadily in recent decades. In 2008, off-farm income sources account for over 93% of the national average farm household income, compared with less than 7% from farming activities (**Figure 10**).

The share of income from farming increases with farm size (as measured by gross sales). "Large" commercial farm households (farms with annual sales between $250,000 and $499,999), on average, obtained 60% of their total household income from farming activities in 2007, while "very large" family farms (farms with annual sales in excess of $500,000) obtained nearly 80% of household income on-farm.[9] These two classes of farms represented slightly less than 8% of family farms. Intermediate family farms (farms with annual sales in excess of $100,000 but less than $250,000) represented about 28% of family farms and obtained about 32% of household income from on-farm sources. The remaining 64% of family farms are classified as rural residence farms and either receive little or no income from farm sources or have a total income level that qualifies them as limited-resource farms.

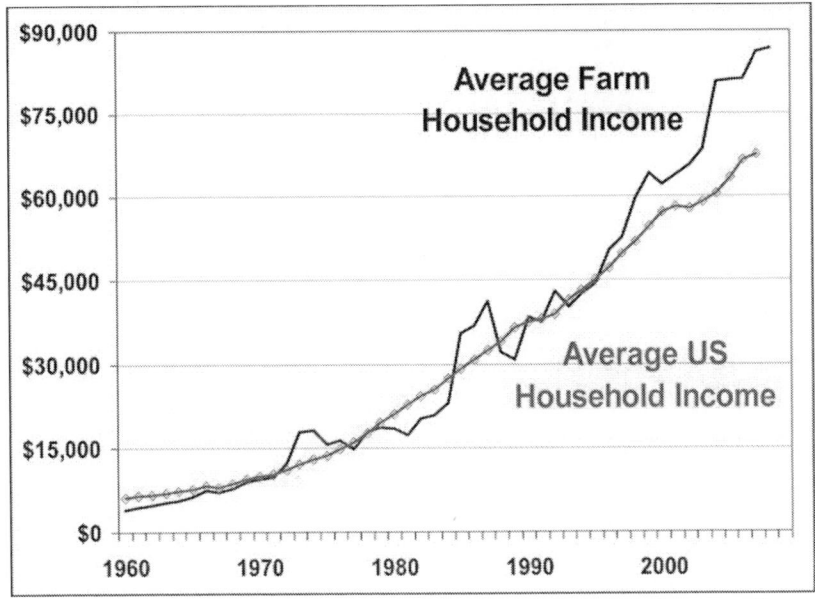

Source: USDA, Economic Research Service, "2008 Farm Income Forecast," November 25, 2008, at http://www.ers.usda.gov/Briefing/FarmIncome/.
Note: 2008 is projected.

Figure 11. Comparison of Farm to U.S. Average Household Income

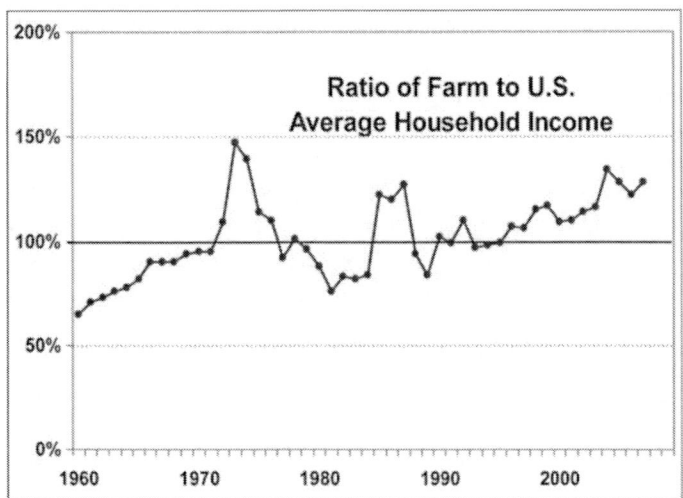

Source: USDA, Economic Research Service, "2008 Farm Income Forecast," November 25, 2008, at http://www.ers.usda.gov/Briefing/FarmIncome/.
Note: 2008 is projected.

Figure 12. Ratio of Farm to U.S. Average Household Income

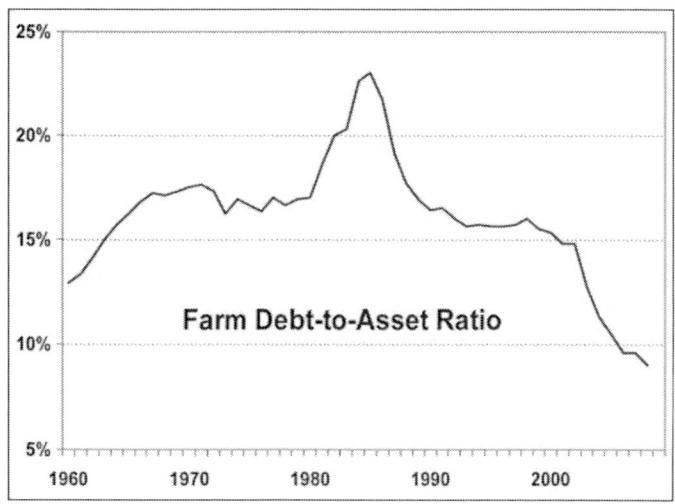

Source: USDA, Economic Research Service, "2008 Farm Income Forecast," November 25, 2008, at http://www.ers.usda.gov/Briefing/FarmIncome/.
Note: 2008 is projected.

Figure 13. U.S. Farm Debt-to-Asset Ratio Since 1960

Over the past decade, farm household incomes have surged ahead of average U.S. household incomes (**Figure 11** and **Figure 12**). In 2007 (the last year for which comparable data were available), the average farm household income of $86,233 was nearly 28% higher than the average U.S. household income (**Table 2**).

FARM ASSET VALUES AND DEBT

Farm asset values — which reflect farm investors' and lenders' expectations about long-term profitability of farm sector investments — are projected up nearly 7% in 2008 to a record $2,359 billion, on the strength of continued growth in real estate values. Farm debt is projected nearly flat at $211.7 billion in 2008, up only $0.2 billion from the previous year's record. As a result, farm equity (or net worth, defined as asset value minus debt) is projected at a record $2,147 billion, while the farm debt-to-asset ratio in 2008 is expected to decline to a 49-year low of 9.0%. The U.S. farm debt-to-asset ratio peaked in 1985 at 23%.

End Notes

[1] For a more detailed discussion and presentation of statistics on the issues discussed in this report, see Agricultural Income and Finance Outlook, AIS-86, ERS, USDA, December 2008.
[2] For more information see CRS Report RL32712, Agriculture-Based Renewable Energy Production, by Randy Schnepf.
[3] USDA, World Agricultural Outlook Board, World Agricultural Supply and Demand Estimates, Jan. 12, 2008.
[4] For more information on commodity programs, see CRS Report RL34594, Farm Commodity Programs *in the 2008 Farm Bill*, by Jim Monke.
[5] For more information, see CRS Report RL3 1095, *Emergency Funding for Agriculture: A Brief History of Supplemental Appropriations, FY1989-FY2009*, by Ralph M. Chite; and CRS Report RS21212, *Agricultural Disaster Assistance*, by Ralph M. Chite.

CHAPTER SOURCES

The following chapters have been previously published:

Chapter 1 – This is an edited, excerpted and augmented edition of a United States Department of Agriculture, Economic Research Service, Economic Information Bulletin Number 42, dated December 2008.

Chapter 2 – This is an edited, excerpted and augmented edition of a United States Congressional Research Service publication, Report Order Code R40152, dated January 16, 2009.

INDEX

A

adjustment, 10
adult, 65
age, 13, 14, 34, 35, 36, 66, 68
agricultural, 2, 6, 10, 12, 13, 14, 22, 44, 46, 47, 51, 57, 59, 67, 72, 73
agricultural sector, x, xi, 72
agriculture, 3, 5, 6, 11, 12, 13, 14, 56, 58, 59, 65, 67, 80
AMS, 84
analysts, 13
animals, 25
antitrust, 59
aquaculture, 25
ash, 81
assets, 18, 19, 40, 41, 42, 43, 46, 47, 50, 52
authors, 2

B

balance sheet, x, 72
barley, 21, 81
beef, 1, 20, 21, 22, 24, 28, 29, 52, 56, 59, 86
benefits, 31, 51, 72, 89

biofuels, 85
birds, 7
bison, 25
breeding, 53
broilers, 7, 76, 84
buildings, 21
Bureau of the Census, 64
business management, 56

C

capital consumption, 88
capital gains, 45
capital intensive, 51
cash flow, 73
cattle, 7, 20, 24, 26, 80
Census, 5, 6, 7, 11, 12, 13, 14, 16, 17, 18, 19, 20, 58, 61, 62, 64, 65, 66, 67, 68
Census Bureau, 62
certificate, 79, 89
classes, 10, 15, 18, 22, 34, 35, 40, 46, 50, 51, 54, 66, 67, 92
classification, 43
Clean Water Act, 60
Cochrane, 7, 61
commodity, 3, 18, 20, 21, 22, 27, 51, 57, 59, 68, 71, 72, 76, 78, 79, 81, 87, 96
Commodity Credit Corporation, 78

compensation, 51
competitive advantage, ix, 1, 3, 55, 56, 57
concentration, 3, 58, 59
confidence, 67
conservation, 21, 79, 87
Conservation Security Program, 22
consumer demand, 54
consumption, 74, 79, 88
contractors, 5, 47, 51, 54, 56, 60
contracts, 4, 6, 7, 41, 47, 51, 52, 54, 55, 60, 68
control, 21, 28, 32, 47, 60
corn, 7, 25, 72, 81, 85
corporations, 1, 4, 15, 29, 32, 41, 56, 60
costs, 60, 71, 80, 88, 91
cotton, 4, 7, 18, 21, 23, 26, 72
counter-cyclical payment, 79, 87, 89
counter-cyclical payments, 79, 87, 89
covering, 18
cows, 7
credit, 72, 85
credit availability, 54
crops, 1, 4, 7, 18, 20, 21, 22, 24, 25, 27, 28, 29, 48, 52, 53, 56, 58, 67, 68, 72, 73, 74, 76, 81, 91
CRP, 22, 25, 87
CRS, 72, 78, 84, 95, 96
cycles, 21

D

dairies, 57
dairy, 7, 20, 21, 22, 24, 26, 27, 28, 48, 56, 57, 58, 59, 69, 72, 80
dairy products, 20
debt, 28, 40, 41, 43, 68, 72, 73, 95
debts, 29
decisions, 33
deficiency, 22, 79, 87, 89
definition, 10, 32, 41, 45
demographic characteristics, 34

Department of Agriculture, 1, 5, 61, 62, 63, 64, 67, 73, 97
Department of Commerce, 10, 61, 64
Department of Justice, 59, 64
depreciation, 79
destruction, 72
deviation, 81
disaster, 22, 71, 87, 89
disaster assistance, 89
distribution, 3, 9, 16, 17, 26, 28
domestic demand, 80

E

earnings, 46
economic activity, 73
economies of scale, 24, 57, 60, 69
education, 37
educational attainment, 34
electricity, 88
elk, 25
energy, 72, 91
Energy Independence and Security Act, 85
energy markets, 72
environment, 21
Environmental Protection Agency, 59
Environmental Quality Incentives Program, 22
EPA, 59
EQIP, 22
equity, 40, 43, 95
estates, 32
ethanol, 80, 85
expenditures, 87

F

family, 1, 2, 4, 6, 15, 31, 32, 33, 35, 47, 56, 60, 73, 92
family farms, 1, 3, 5, 24, 25, 26, 27, 46, 69

Index

family members, 35
Farm Act, 79
Farm Bill, 78, 84, 96
farm bills, 87
farm size, 6, 24, 26, 34, 35, 68, 92
farmers, 18, 58, 66, 74, 80, 84
farming, 3, 6, 14, 21, 34, 36, 45, 56, 60, 66, 92
farmland, 7, 27
Federal Trade Commission, 59, 64
fee, 5, 7, 51
feeding, 60
fees, 7, 68, 77
fertilizers, 88, 91
fiber, 2
field crops, 24, 72, 76, 81
financial crisis, x, 72
financial performance, 2, 41
firms, 31, 47, 58
flexibility, 29, 79, 87
flow, 60, 73
focusing, 6
food, 2, 4, 18, 81
formal education, 34
fruits, 23, 25, 26, 72
fuel, 88
full capacity, 60
funds, 41, 73

G

generation, 4, 35, 38
government, 59, 77, 89
government payments, 58, 66
grain, 7, 25, 27, 51, 52, 53, 57
grains, 4, 18, 22, 25, 53, 60, 80, 84
grazing, 31, 33
greenhouse, 23, 25, 26, 73
groups, 44
growth, 3, 10, 57, 60, 71, 89, 95
growth rate, 89
guidelines, 59

H

handling, 21
harvest, 18, 21
harvesting, 58
hog, 7, 21, 23, 26, 58, 59, 69, 80
hogs, 1, 7, 20, 22, 27, 56, 59, 86
honey, 25
Horizontal Merger Guidelines, 58, 64
horses, 25
household, 2, 6, 32, 35, 44, 46, 68, 74, 81, 92, 95
household income, 2, 44, 46, 92, 95
households, 5, 6, 35, 43, 45, 46, 66, 68, 92

I

ideal, 32, 72
identification, 65
incentive, 35
incentives, 85
inclusion, 78
income, 2, 5, 22, 40, 41, 42, 43, 45, 46, 68, 71, 72, 73, 74, 76, 77, 78, 79, 81, 87, 89, 92, 95
incomes, 95
independence, 61
indicators, 73
industry, 14, 58, 59, 80, 85
inflation, 75, 90
insurance, 77
Internal Revenue Service, 29, 64
inventories, 53, 74
investment, 72
investors, 95

J

Jun, 83

L

labor, 4, 6, 21, 35, 43, 47, 48, 49, 50, 51, 53, 56, 57, 60, 68, 79
land, 7, 18, 28, 47, 66
large-scale, 3, 21, 56, 57
law, 15, 29, 31, 85
LDP, 79
legislation, 59, 89
lenders, 95
lettuce, 7, 58, 59
lifestyle, 45
limitations, 13
limited liability, 6, 29
links, 5, 6, 12, 13, 65, 67
livestock, 5, 7, 21, 25, 48, 54, 55, 59, 67, 68, 71, 72, 74, 76, 77, 80, 84, 86, 87, 88, 91
livestock, 52, 76, 78, 86
loans, 78
local government, 59
location, 15
loss assistance payments, 87
losses, 45

M

machinery, 47, 48, 53
magnetic, iv
maintenance, 88
management, 6, 29, 34, 43, 56, 68
manufacturing, 56
manure, 59
market, 3, 58, 59, 61, 67, 72, 73, 80, 86, 87, 89
market prices, 80, 86, 89
market structure, 73
market value, 67
marketing, 4, 6, 22, 41, 47, 51, 55, 59, 72, 73, 79, 85, 87, 88, 89
marketplace, 32
markets, 60, 72

marriage, 32
measures, 40, 41, 42, 68
meat, 72
median, 26, 27, 28, 43
mergers, 59
milk, 1, 7, 20, 22, 24, 25, 79, 84, 86, 87
model, 56
movement, 21

N

net farm income, 31, 32, 33, 52, 53, 54, 55, 59
net income, 45
normal, 67
nutrients, 60
nutrition, 21
nuts, 23, 25, 26, 72

O

observations, 25, 31, 37, 50, 55, 67, 68
oil, 81, 83, 84
oils, 88
oligopolies, 58
operator, 1, 2, 4, 6, 7, 31, 32, 33, 34, 35, 38, 39, 40, 45, 46, 48, 51, 56, 61, 65, 67, 68, 73, 78, 79, 88
operators, 34, 39, 45
order, 47
outsourcing, 56
ownership, 28, 32, 33, 56, 59, 60

P

partnership, 31
partnerships, 15, 29, 31
peanuts, 21, 25
pesticides, 88
petroleum, 88
pollination, 47, 48
positive relation, 57

Index

positive relationship, 57
poultry, 1, 5, 20, 21, 22, 26, 54, 55, 56, 58, 59, 68, 72, 80, 86, 88
power, 3, 58, 59
price changes, 11, 12
prices, 7, 71, 72, 74, 76, 77, 80, 81, 84, 85, 86, 87, 89, 91
Producer Price Index, 11, 12, 16, 17
producers, 51, 57, 58, 60, 91
production costs, 71
profit, 3, 35, 40, 41, 43, 44, 45, 55, 57
profit margin, 3, 36, 40, 41, 43, 44, 45, 55, 57
profitability, 40, 71, 73, 95
profits, 35
program, 18, 20, 21, 22, 71, 78, 79, 87, 89
property, iv, 88
property taxes, 88
protein, 84
proxy, 68

Q

questionnaire, 54, 65, 67

R

rain, 25
range, 15, 40, 46, 73, 84
rate of return, 41
real estate, 50, 53, 95
recession, x, 72
regulations, 59
relationship, 57
relationships, 60
relatives, 1, 31, 32, 33
relief, 87
rent, 28, 48, 74, 88
repair, 88
resources, 15, 47
revenue, 7, 28

rice, 21, 25, 72, 76, 81
risk, 41
Risk Management Agency, 87
runoff, 60
rural, 45, 73, 92

S

school, 34, 37
seed, 7, 88
services, 7
shareholders, 31, 33
shares, 3, 41, 55
sheep, 25
skills, 56
sole proprietor, 15, 29, 31
solvency, 43
solvent, 41, 68
soybean, 7, 84
soybeans, 7, 21, 25, 72, 81
specialization, 7, 24, 26, 27, 36, 58
specialty crop, 7, 28, 58
species, 25
spectrum, 2, 7
spouse, 6, 35, 40, 48, 56, 61, 68
stabilize, 74
standard error, 19, 28, 31, 37, 40, 43, 44, 46, 50, 53
standards, 59
statistics, 51, 73, 95
stock, 53
storage, 47, 73, 88
strategies, 6, 41
strength, 95
sugar, 25
sugar beet, 25
sugar cane, 25
supplemental, 87
suppliers, 56
supply, 51, 56

T

tariff, 85
tax credit, 85
taxation, 29, 45
taxes, 31, 67, 88
technological advancement, 21
technological change, 56
technology, 32
tenants, 28
tenure, 6, 27
time, 13, 14, 15, 32, 35, 66, 73
timing, 74, 89
tobacco, 21, 87
tomato, 7
tourism, 79
transactions, 60, 74
transition, 87
transportation, 21, 88
Treasury, 29, 64
triggers, 72
trusts, 31, 32, 33
turnover, 14

U

United States Department of Agriculture, 5, 61, 62, 63, 64, 67, 73
U.S. Department of Agriculture (USDA), 73
U.S. Department of the Treasury, 29, 64
uncertainty, 72
unemployment, 72

V

values, 46, 67, 72, 75, 90, 95
variability, 21, 74
vegetables, 22, 73
vehicles, 47
volatility, 7

W

wage rate, 61, 68
wages, 68
well-being, 71, 73
wheat, 21, 25, 72, 81
wholesale, 84
wool, 25
workers, 49, 68
workforce, 36